Building Serverless Microservices in Python

A complete guide to building, testing, and deploying microservices using serverless computing on AWS

Richard Takashi Freeman

BIRMINGHAM - MUMBAI

Building Serverless Microservices in Python

Commissioning Editor: Richa Tripathi
Acquisition Editor: Denim Pinto
Content Development Editor: Rohit Kumar Singh
Technical Editor: Romy Dias
Copy Editor: Safis Editing
Project Coordinator: Vaidehi Sawant
Proofreader: Safis Editing
Indexer: Pratik Shirodkar
Graphics: Alishon Mendonsa
Production Coordinator: Deepika Naik

First published: March 2019

Production reference: 2290319

Published by Packt Publishing Ltd.
Livery Place
35 Livery Street
Birmingham
B3 2PB, UK.

ISBN 978-1-78953-529-7

www.packtpub.com

`mapt.io`

Mapt is an online digital library that gives you full access to over 5,000 books and videos, as well as industry leading tools to help you plan your personal development and advance your career. For more information, please visit our website.

Why subscribe?

- Spend less time learning and more time coding with practical eBooks and Videos from over 4,000 industry professionals

- Improve your learning with Skill Plans built especially for you

- Get a free eBook or video every month

- Mapt is fully searchable

- Copy and paste, print, and bookmark content

Packt.com

Did you know that Packt offers eBook versions of every book published, with PDF and ePub files available? You can upgrade to the eBook version at `www.packt.com` and as a print book customer, you are entitled to a discount on the eBook copy. Get in touch with us at `customercare@packtpub.com` for more details.

At `www.packt.com`, you can also read a collection of free technical articles, sign up for a range of free newsletters, and receive exclusive discounts and offers on Packt books and eBooks.

Contributors

About the author

Richard Takashi Freeman has an M.Eng. in computer system engineering and a PhD in machine learning and natural language processing from the University of Manchester, United Kingdom. His current role is as a lead data engineer and architect, but he is also a data scientist and solutions architect. He has been delivering cloud-based, big data, machine learning, and data pipeline serverless and scalable solutions for over 14 years, and has spoken at numerous leading academic and industrial conferences, events, and summits.

He has written various technical blog posts, and has acquired in excess of three years' serverless production experience in connection with large-scale, consumer-facing websites and products.

Packt is searching for authors like you

If you're interested in becoming an author for Packt, please visit `authors.packtpub.com` and apply today. We have worked with thousands of developers and tech professionals, just like you, to help them share their insight with the global tech community. You can make a general application, apply for a specific hot topic that we are recruiting an author for, or submit your own idea.

Table of Contents

Preface

This book will give you a very good understanding of microservices and serverless computing, as well as their benefits and drawbacks compared to existing architectures. You will gain an appreciation for the power of deploying a full serverless stack, not only when you save money in terms of running costs, but also in terms of support maintenance and upgrading. This effectively allows your company to go to market a lot quicker with any new products, and beat your competitors in the process. You will also be able to create, test, and deploy a scalable, serverless microservice where the costs are paid per usage and not as per their up time. In addition, this will allow you to autoscale based on the number of requests, while security is natively built in and supported by AWS. So, now that we know what lies ahead, let's jump right into the book.

Who this book is for

If you are a developer with basic knowledge of Python and want to learn how to build, test, deploy, and secure microservices, then this book is for you. No prior knowledge of building microservices is required.

What this book covers

Chapter 1, *Serverless Microservice Architectures and Patterns*, provides an overview of monolithic and microservice architectures. You will learn about design patterns and principles and how they relate to serverless microservices.

Chapter 2, *Creating Your First Serverless Data API*, discusses security and its importance. We will discuss IAM roles and get an overview of policies and some of the security concepts and principles involved in securing your serverless microservices, specifically regarding Lambda, API Gateway, and DynamoDB.

Chapter 3, *Deploying Your Serverless Stack*, shows you how to deploy all that infrastructure using only code and configuration. You will learn about different deployment options.

Chapter 4, *Testing Your Serverless Microservice*, covers the concept of testing. We will explore many types of testing, from unit tests with mocks, integration testing with Lambda and API Gateway, debugging a Lambda locally, and making a local endpoint available, to load testing.

Chapter 5, *Securing Your Microservice*, covers important topics on how to make your microservices secure.

To get the most out of this book

Some prior programming knowledge will be helpful.

All the other requirements will be mentioned at the relevant points in the respective chapters.

Download the example code files

You can download the example code files for this book from your account at www.packt.com. If you purchased this book elsewhere, you can visit www.packt.com/support and register to have the files emailed directly to you.

You can download the code files by following these steps:

1. Log in or register at www.packt.com.
2. Select the **SUPPORT** tab.
3. Click on **Code Downloads & Errata**.
4. Enter the name of the book in the **Search** box and follow the onscreen instructions.

Once the file is downloaded, please make sure that you unzip or extract the folder using the latest version of:

- WinRAR/7-Zip for Windows
- Zipeg/iZip/UnRarX for Mac
- 7-Zip/PeaZip for Linux

The code bundle for the book is also hosted on GitHub at `https://github.com/PacktPublishing/Building-Serverless-Microservices-in-Python`. In case there's an update to the code, it will be updated on the existing GitHub repository.

We also have other code bundles from our rich catalog of books and videos available at `https://github.com/PacktPublishing/`. Check them out!

Download the color images

We also provide a PDF file that has color images of the screenshots/diagrams used in this book. You can download it here: `http://www.packtpub.com/sites/default/files/downloads/9781789535297_ColorImages.pdf`.

Conventions used

There are a number of text conventions used throughout this book.

`CodeInText`: Indicates code words in text, database table names, folder names, filenames, file extensions, pathnames, dummy URLs, user input, and Twitter handles. Here is an example: "Here, you can see that we have `EventId` as resource `1234`, and a `startDate` parameter formatted in the `YYYYMMDD` format."

A block of code is set as follows:

```
"phoneNumbers": [
  {
    "type": "home",
    "number": "212 555-1234"
  },
  {
```

When we wish to draw your attention to a particular part of a code block, the relevant lines or items are set in bold:

```
{
  "firstName": "John",
  "lastName": "Smith",
  "age": 27,
  "address": {
```

Any command-line input or output is written as follows:

```
$ cd /mnt/c/
```

Bold: Indicates a new term, an important word, or words that you see on screen. For example, words in menus or dialog boxes appear in the text like this. Here is an example: "In the DynamoDB navigation pane, choose **Tables** and choose **user-visits**."

 Warnings or important notes appear like this.

 Tips and tricks appear like this.

Get in touch

Feedback from our readers is always welcome.

General feedback: If you have questions about any aspect of this book, mention the book title in the subject of your message and email us at customercare@packtpub.com.

Errata: Although we have taken every care to ensure the accuracy of our content, mistakes do happen. If you have found a mistake in this book, we would be grateful if you would report this to us. Please visit www.packt.com/submit-errata, selecting your book, clicking on the Errata Submission Form link, and entering the details.

Piracy: If you come across any illegal copies of our works in any form on the internet, we would be grateful if you would provide us with the location address or website name. Please contact us at copyright@packt.com with a link to the material.

If you are interested in becoming an author: If there is a topic that you have expertise in, and you are interested in either writing or contributing to a book, please visit authors.packtpub.com.

Reviews

Please leave a review. Once you have read and used this book, why not leave a review on the site that you purchased it from? Potential readers can then see and use your unbiased opinion to make purchase decisions, we at Packt can understand what you think about our products, and our authors can see your feedback on their book. Thank you!

For more information about Packt, please visit `packt.com`.

1
Serverless Microservices Architectures and Patterns

Microservices architectures are based on service. You could think of the microservices as a lightweight version of SOA but enriched with more recent architectures, such as the event-driven architecture, where an event is defined as a state of change that's of interest. In this chapter, you will learn about the monolithic multi-tier architecture and monolithic **service-oriented architecture (SOA)**. We will discuss the benefits and drawbacks of both architectures. We will also look at the microservices background to understand the rationale behind its growth, and compare different architectures.

We will cover the design patterns and principles and introduce the serverless microservice integration patterns. We then cover the communication styles and decomposition microservice patterns, including synchronous and asynchronous communication.

You will then learn how serverless computing in AWS can be used to quickly deploy event-driven computing and microservices in the cloud. We conclude the chapter by setting up your serverless AWS and development environment.

In this chapter we will cover the following topics:

- Understanding different architecture types and patterns
- Virtual machines, containers, and serverless computing
- Overview of microservice integration patterns
- Communication styles and decomposition microservice patterns
- Serverless computing in AWS
- Setting up your serverless environment

Understanding different architecture types and patterns

In this section, we will discuss different architectures, such as monolithic and microservices, along with their benefits and drawbacks.

The monolithic multi-tier architecture and the monolithic service-oriented architecture

At the start of my career, while I was working for global fortune 500 clients for Capgemini, we tended to use multi-tier architecture, where you create different physically separate layers that you can update and deploy independently. For example, as shown in the following three-tier architecture diagram, you can use **Presentation**, **Domain logic**, and **Data Storage** layers:

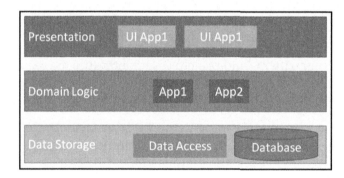

In the **presentation layer**, you have the user interface elements and any presentation-related applications. In **domain logic**, you have all the business logic and anything to do with passing the data from the presentation layer. Elements in the domain logic also deal with passing data to the **storage or data layer**, which has the data access components and any of the database elements or filesystem elements. For example, if you want to change the database technology from SQL Server to MySQL, you only have to change the data-access components rather than modifying elements in the presentation or domain-logic layers. This allows you to decouple the type of storage from presentation and business logic, enabling you to more readily change the database technology by swapping the data-storage layer.

A few years later at Capgemini, we implemented our clients' projects using SOA, which is much more granular than the multi-tier architecture. It is basically the idea of having standardized service contracts and registry that allows for automation and abstraction:

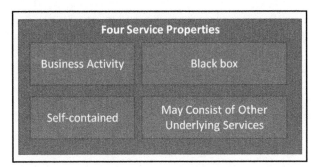

There are four important service properties related to SOA:

- Each service needs to have a clear business activity that is linked to an activity in the enterprise.
- Anybody consuming the service does not need to understand the inner workings.
- All the information and systems are self-contained and abstracted.
- To support its composability, the service may consist of other underlying services

Here are some important SOA principles:

- Standardized
- Loosely coupled
- Abstract
- Stateless
- Granular
- Composable
- Discoverable
- Reusable

The first principle is that there is a **standardized** service contract. This is basically a communication agreement that's defined at the enterprise level so that when you consume a service, you know exactly which service it is, the contract for passing in messages, and what you are going to get back. These services are **loosely coupled**. That means they can work autonomously, but also you can access them from any location within the enterprise network. They also offer an **abstract** version, which means that these services are a black box where the inner logic is actually hidden away, but also they can work independently of other services.

Some services will also be **stateless**. That means that, if you call a service, passing in a request, you will get a response and you would also get an exception if there is a problem with the service or the payload. Granularity is also very important within SOA. The service needs to be **granular** enough that it's not called inefficiently or many times. So, we want to normalize the level and the granularity of the service. Some services can be decomposed if they're being reused by the services, or services can be joined together and normalized to minimize redundancy. Services also need to be **composable** so you can merge them together into larger services or split them up.

There's a standardized set of contracts, but the service also needs to be **discoverable**. Discoverable means that there is a way to automatically discover what service is available, what endpoints are available, and a way to interpret them. Finally, the reasonable element, **reuse** is really important for SOA, which is when the logic can be reused in other parts of the code base.

Benefits of monolithic architectures

In SOA, the architecture is loosely coupled. All the services for the enterprise are defined in one repository. This allows us to have good visibility of the services available. In addition, there is a global data model. Usually, there is one data store where we store all the data sources and each individual service actually writes or reads to it. This allows it to be centralized at a global level.

Another benefit is that there is usually a small number of large services, which are driven by a clear business goal. This makes them easy to understand and consistent for our organization. In general, the communication between the services is decoupled via either smart pipelines or some kind of middleware.

Drawbacks of the monolithic architectures

The drawback of the monolithic architecture is that there is usually a single technology stack. This means the application server or the web server or the database frameworks are consistent throughout the enterprise. Obsolete libraries and code can be difficult to upgrade, as this is dependent on a single stack and it's almost like all the services need to be aligned on the same version of libraries.

Another drawback is that the code base is usually very large on a single stack stack, which means that there are long build times and test times to build and deploy the code. The services are deployed on a single or a large cluster of application servers and web servers. This means that, in order to scale, you need to scale the whole server, which means there's no ability to deploy and scale applications independently. To scale out an application, you need to scale out the web application or the application server that hosts the application.

Another drawback is that there's generally a middleware orchestration layer or integration logic that is centralized. For example, services would use the **Business Process Management (BPM)** framework to control the workflow, you would use an **Enterprise Service Bus (ESB)**, which allows you to do routing your messages centrally, or you'd have some kind of middleware that would deal with the integration between the services themselves. A lot of this logic is tied up centrally and you have to be very careful not to break any inter-service communication when you're changing the configuration of that centralized logic.

Overview of microservices

The term microservice arose from a workshop in 2011, when different teams described an architecture style that they used. In 2012, Adrien Cockcroft from Netflix actually described microservice as a fine-grained SOA who pioneered this fine-grained SOA at web scale.

For example, if we have sensors on an **Internet of Things (IoT)** device, if there's a change of temperature, we would emit an event as a possible warning further downstream. This is what's called **event-stream processing** or **complex-event processing**. Essentially, everything is driven by events throughout the whole architecture.

The other type of design used in microservices is called **domain-driven design** (DDD). This is essentially where there is a common language between the domain experts and the developers. The other important component in DDD is the **bounded context**, which is where there is a strict model of consistency that relies in its bounds for each service. For example, if it's a service dealing with customer invoicing, that service will be the central point and only place where customer invoicing can be processed, written to, or updated. The benefits are that there won't be any confusion around the responsibilities of data access with systems outside of the bounded context.

You could think of microservice as centered around a REST endpoint or application programming interface using JSON standards. A lot of the logic could be built into the service. This is what is called a **dumb pipeline** but a smart endpoint, and you can see why in the diagram. We have a service that deals with customer support, as follows:

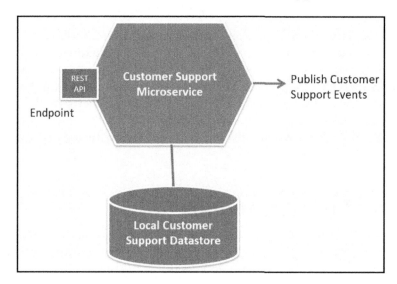

For example, the endpoint would update customer support details, add a new ticket, or get customer support details with a specific identifier. We have a local customer support data store, so all the information around customer support is stored in that data store and you can see that the microservice emits customer-support events. These are sent out on a publish-subscribe mechanism or using other publishing-event frameworks, such as **Command Query Responsibility Segregation** (CQRS). You can see that this fits within the bounded context. There's a single responsibility around this bounded context. So, this microservice controls all information around customer support.

Benefits and drawbacks of microservice architectures

The bounded context, and the fact that this is a very small code base, allow you to build very frequently and deploy very frequently. In addition, you can scale these services independently. There's usually one application server or web server per microservice. You can obviously scale it out very quickly, just for the specific service that you want to. In addition, you can have frequent builds that you test more frequently, and you can use any type of language, database, or web app server. This allows it to be a polygon system. The bounded context is a very important as you can model one domain. Features can be released very quickly because, for example, the customer services microservice could actually control all changes to the data, so you can deploy these components a lot faster.

However, there are some drawbacks to using a microservices architecture. First, there's a lot of complexity in terms of distributed development and testing. In addition, the services talk a lot more, so there's more network traffic. Latency and networks become very important in microservices. The DevOps team has to maintain and monitor the time it takes to get a response from another service. In addition, the changing of responsibilities is another complication. For example, if we're splitting up one of the bounded contexts into several types of sub-bounded context, you need to think about how that works within teams. A dedicated DevOps team is also generally needed, which is essentially there to support and maintain much larger number of services and machines throughout the organization.

SOA versus microservices

Now that we have a good understanding of both, we will compare the SOA and microservices architectures. In terms of the communication itself, both SOA and microservices can use synchronous and asynchronous communication. SOA typically relied on **Simple Object Access Protocol (SOAP)** or web services. Microservices tend to be more modern and widely use **REpresentational State Transfer (REST) Application Programming Interfaces (APIs)**.

We will start with the following diagram, which compares SOA and microservices:

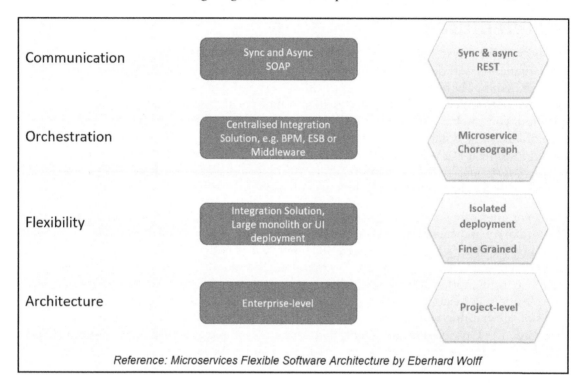

Communication	Sync and Async SOAP	Sync & async REST
Orchestration	Centralised Integration Solution, e.g. BPM, ESB or Middleware	Microservice Choreograph
Flexibility	Integration Solution, Large monolith or UI deployment	Isolated deployment Fine Grained
Architecture	Enterprise-level	Project-level

Reference: Microservices Flexible Software Architecture by Eberhard Wolff

The orchestration is where there's a big differentiation. In SOA, everything is centralized around a BPM, ESB, or some kind of middleware. All the integration between services and data flowing is controlled centrally. This allows you to configure any changes in one place, which has some advantages.

The microservices approach has been to use a more choreography-based approach. This is where an individual service is smarter, that is, a smart endpoint but a dumb pipeline. That means that the services know exactly who to call and what data they will get back, and they manage that process within the microservice. This gives us more flexibility in terms of the integration for microservices. In the SOA world or the three-tier architecture, there's less flexibility as it's usually a single code base and the integration is a large set of monolith releases and deployments of user interface or backend services. This can limit the flexibility of your enterprise. For microservices, however, these systems are much smaller and can be deployed in isolation and much more fine-grained.

Finally, on the architecture side, SOA works at the enterprise level, where we would have an enterprise architect or solutions architect model and control the release of all the services in a central repository. Microservices are much more flexible. Microservices talked about working at the project level where they say the team is only composed of a number of developers or a very small number of developers that could sit around and share a pizza. So, this gives you much more flexibility to make decisions rapidly at the project level, rather than having to get everything agreed at the enterprise level.

Virtual machines, containers, and serverless computing

Now that we have a better understanding of the monolithic and microservice architectures, let's look at the **Amazon Web Service (AWS)** building blocks for creating serverless microservices.

But first we'll cover virtual machines, containers, and serverless computing, which are the basic building blocks behind any application or service hosted in the public cloud.

Virtual machines are the original offering in the public cloud and web hosting sites, containers are lightweight standalone images, and serverless computing is when the cloud provider fully manages the resources. You will understand the benefits and drawbacks of each approach and we will end on a detailed comparison of all three.

Virtual machines

In traditional data centers, you would have to buy or lease physical machines and have spare capacity to deal with additional web or user traffic. In the new world, virtual machines were one of the first public cloud offerings. You can think of it as similar to physical boxes, where you can install an operating system, remotely connect via SSH or RDP, and install applications and services. I would say that virtual machines have been one of the key building blocks for start-up companies to be successful. It gave them the ability to go to market with only small capital investments and to scale out with an increase in their web traffic and user volumes. This was something that previously only large organizations could afford, given the big upfront costs of physical hardware.

The advantages of virtual machines are the pay per usage, choice of instance type, and dynamic allocation of storage, giving your organization full flexibility to rent hardware within minutes rather than wait for physical hardware to be purchased. Virtual machines also provides security, which is managed by the cloud provider. In addition, they provide multi-region auto-scaling and load balancing, again managed by the cloud provider and available almost at the click of a button. There are many virtual machines available, for example, Amazon EC2, Azure VMs, and Google Compute Engine.

However, they do have some drawbacks. The main drawback is that it takes a few minutes to scale. So, any machine that needs to be spun up takes a few minutes, making it impossible most to scale quickly upon request. There is also an effort in terms of configuration where the likes of Chef or Puppet are required for configuration management. For example, the operating system needs to be kept up to date.

Another drawback is that you still need to write the logic to poll or subscribe to other managed services, such as streaming analytics services. In addition, you still pay for idle machine time. For example, when your services are not running, the virtual machines are still up and you're still paying for that time even if they're not being actively used.

Containers

The old way with virtual machines was to deploy applications on a host operating system with configuration-management tools such as Chef or Puppet. This has the advantage of managing the application artifacts' libraries and life cycles with each other and trying to operate specific operating systems, whether Linux or Windows. Containers came out of this limitation with the idea of shipping your code and dependencies into a portable container where you have full operating-system-level virtualization. You essentially have better use of the available resources on the machine.

These containers can be spun up very fast and they are essentially immutable, that is, the OS, library versions, and configurations cannot be changed. The basic idea is that you ship the code and dependencies in this portable container and the environments can be recreated locally or on a server by a configuration. Another important aspect is the orchestration engine. This is the key to managing containers. So, you'd have Docker images that will be managed, deployed, and scaled by Kubernetes or Amazon **EC2 container service (ECS)**.

The drawbacks are that these containers generally scale within seconds, which is still too slow to actually invoke a new container per request. So, you'd need them to be pre-warmed and already available, which has a cost. In addition, the cluster and image configuration does involve some DevOps effort.

Recently AWS introduced AWS Fargate and **Elastic Kubernetes Service (EKS)**, which have helped to relieve some of this configuration-management and support effort, but you would still need a DevOps team to support them.

The other drawback is that there's an integration effort with the managed services. For example, if you're dealing with a streaming analytics service, you still need to write the polling and subscription code to pull the data into your application or service.

Finally, like with virtual machines, you still pay for any containers that are running even if the Kubernetes assists with this. They can run on the EC2 instance, so you'll still need to pay for that actual machine through a running time even if it's not being used.

Serverless computing

You can think of service computing as focusing on business logic rather than on all the infrastructure-configuration management and integration around the service. In serverless computing, there are still servers, it's just that you don't manage the servers themselves, the operating system, or the hardware, and all the scalability is managed by the cloud provider. You don't have access to the raw machine, that is, you can't SSH onto the box.

The benefits are that you can really focus on the business logic code rather than any of the infrastructure or inbound integration code, which is the the business value you are adding as an organization for your customers and clients.

In addition, the security is managed by the cloud provider again, auto-scaling and the high availability options also managed by the cloud provider. You can spin up more instances dynamically based on the number of requests, for example. The cost is per execution time not per idle time.

There are different public cloud serverless offerings. Google, Azure, AWS, and Alibaba cloud have the concept of **Functions as a Service (FaaS)**. This is where you deploy your business logic code within a function and everything around it, such as the security and the scalability, is managed by the cloud provider.

The drawback is that these are stateless, which means they have a very short lifetime. After the few minutes are over, any state maintained within that function is lost, so it has to be persisted outside. It's not suitable for a long-running processes. It does have a limited instance type and a duration too. For example, AWS Lambdas have a duration of 15 minutes before they are terminated. There's also constraints on the actual size of the external libraries or any custom libraries that you package together, since these lambdas need to be spun up very quickly.

Comparing virtual machines, containers, and serverless

Let's compare **Infrastructure as a Service (IaaS)**, **Containers as a Service (CaaS)**, and **Functions as a Service (FaaS)**. Think of IaaS as the virtual machine, CaaS as pool of Docker containers and FaaS an example will be Lambda functions. This is a comparison between IaaS, CaaS, and FaaS:

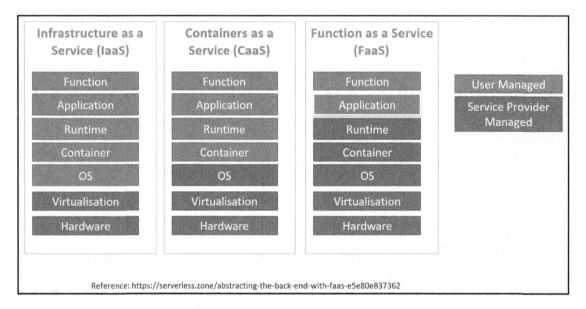

Reference: https://serverless.zone/abstracting-the-back-end-with-faas-e5e80e837362

The green elements are managed by the user, and the blue elements are managed by the cloud service provider. So, on the left, you can see that IaaS, as used with virtual machines, have a lot of the responsibility on the user. In CaaS, the operating-system level is managed by the provider, but you can see that the container and the runtime are actually managed by the user. And, finally on the right, FaaS, you can see the core business logic code and application configuration is managed by the user.

So, how do you choose between AWS Lambda containers and EC2 instances in the AWS world? Check out the following chart:

	Virtual Machines / Amazon EC2 Instance	Containers / Amazon ECS	AWS Lambda Functions
Infrastructure	You configure, maintain and built for HA	You configure, maintain and built for HA	Request-driven and AWS manages HA
Instance Flexibility	Choose instance type, OS, etc.	Choose instance type, OS, etc.	Default hardware & OS, no maintenance
Scaling	Provisioning instances and configure auto-scaling	Provisioning containers and configure auto-scaling	Implicit, based on requests
Launch / Lifetime	Minutes / Live for weeks	Seconds / Live for minutes or hours	Milliseconds to few seconds / Live for seconds
State	State or stateless	State or stateless	Stateless
AWS Integration	Custom	Custom	Built-in triggers SNS, Kinesis Stream, S3, API Gateway etc.
Pricing	EC2 per second and spot	Containers on EC2 clusters per second and spot	Per 100ms, invocation and RAM

If we compare virtual machines against the containers and Lambda functions on the top row, you can see that there is some configuration effort required in terms of the maintenance, building it for high availability, and management. For the Lambda functions, this is actually done on a pre-request basis. That is, it's request-driven. AWS will spin up more lambdas if more traffic hits your site to make it **highly available (HA)**, for example.

In terms of flexibility, you have full access in virtual machines and containers, but with AWS Lambda, you have default hardware, default operating system, and no **graphics processing units (GPU)** available. The upside is that there is no upgrade or maintenance required on your side for Lambdas.

In terms of scalability, you need to plan ahead for virtual machines and containers. You need to provision the containers or instances and decide how you are going to scale. In AWS Lambda functions, scaling is implicit based on the number of requests or data volumes, as you natively get more or fewer lambdas executing in parallel.

The launch of virtual machines is usually in minutes and they can stay on perhaps for weeks. Containers can spin up within seconds and can stay on for minutes or hours before they can be disposed of. Lambda functions, however, can spin up in around 100 milliseconds and generally live for seconds or maybe a few minutes.

In terms of state, virtual machines and containers can maintain state even if it's generally not best practice for scaling. Lambda functions are always stateless, when they terminate their execution, anything in memory is disposed of, unless it's persisted outside in a DynamoDB table or S3 bucket, for example.

Custom integration with AWS services is required for virtual machines and Docker containers. In Lambda functions, however, event sources can push data to a Lambda function using built-in integration with the other AWS services, such as Kinesis, S3, and API Gateway. All you have to do is subscribe the Lambda event source to a Kinesis Stream and the data will get pushed to your Lambda with its business logic code, which allows you to decide how you process and analyze that data. However, for EC2 virtual machines and ECS containers, you need to build that custom inbound integration logic using the AWS SDK, or by some other means.

Finally, in terms of pricing, EC2 instances are priced per second. They also have a spot instance that uses market rates, which is lot cheaper than on-demand instances. The same goes for containers, except that you can have many containers on one EC2 instance. This makes better use of resources and is a lot cheaper, as you flexibility to spread different containers among the EC2 instances. For AWS Lambda functions, the pricing is per 100 milliseconds, invocation number, and the amount of **random-access memory (RAM)** required.

Overview of microservice integration patterns

In this section, we'll discuss design patterns, design principles, and how microservice architectural patterns relate to traditional microservice patterns and can be applied to serverless microservices. These topics will help you gain an overview of different integration patterns.

Design patterns

Patterns are reusable blueprints that are a solution to a similar problem others have faced, and that have widely been reviewed, tested, and deployed in various production environments.

Following them means that you will benefit from best practices and the wisdom of the technical crowd. You will also speak the same language as other developers or architects, which allows you to exchange your ideas much faster, integrate with other systems more easily, and run staff handovers more effectively.

Why are patterns useful?

Useful applications almost never exist in isolation. They are almost always integrated in a wider ecosystem, which is especially true for microservices. In other words, the integration specification and requirements need to be communicated and understood by other developers and architects.

When using patterns, you have a common language that is spoken among the technical crowds, allowing you to be understood. It's really about better collaborating, working with other people, exchanging ideas, and working out how to solve problems.

The main aim of patterns is to save you time and effort when implementing new services, as you have a standard terminology and blueprint to build something. In some cases, they help you avoid pitfalls as you can learn from others' experience and also apply the best practices, software, design patterns, and principles.

Software design patterns and principles

Your will probably be using **object-oriented** (**OO**) or functional programming in your microservices or Lambda code, so let's briefly talk about the patterns linked to them.

In OO programming, there are many best practice patterns or principles you can use when coding, such as GRASP or SOLID. I will not go into too much depth as it would take a whole book, but I would like to highlight some principles that are important for microservices:

- **SOLID**: This has five principles. One example is the **Single Responsibility Principle (SRP)**, where you define classes that each have a single responsibility and hence a single reason for change, reducing the size of the services and increasing their stability.

- **Package cohesion**: For example, common closure-principle classes that change together belong together. So when a business rule changes, developers only need to change code in a small number of packages.
- **Package coupling**: For example, the acyclic dependencies principle, which states that dependency graphs of packages or components should have no cycles.

Let's briefly go into some of the useful design patterns for microservice:

- **Creational patterns**: For example, the factory method creates an instance of several derived classes.
- **Structural patterns**: For example, the decorator adds additional responsibilities to an object dynamically.
- **Behavioral patterns**: For example, the command pattern encapsulates a request as an object, making it easier to extract parameters, queuing, and logging of requests. Basically, you decouple the parameter that creates the command from the one that executes it.
- **Concurrency patterns**: For example, the reactor object provides an asynchronous interface to resources that must be handled synchronously.

Depending on you coding experience, you may be familiar with these. If not, it's worth reading about them to improve you code readability, management, and stability, as well as your productivity. Here are some references where you can find out more:

- *SOLID Object-Oriented Design*, Sandi Metz (2009)
- *Design Patterns: Elements of Reusable Object-Oriented Software*, Erich Gamma, Richard Helm, Ralph Johnson, John Vlissides (1995)
- *Head First Design Patterns*, Eric T Freeman, Elisabeth Robson, Bert Bates, Kathy Sierra (2004)
- *Agile Software Development, Principles, Patterns, and Practices*, Robert C. Martin (2002)

Serverless microservices pattern categories

On top of the software design patterns and principles we just discussed are the microservices patterns. From my experience, there are many microservices patterns that I recommended that are relevant for serverless microservices, as shown in the following diagram:

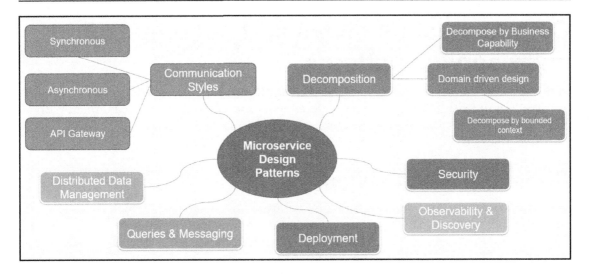

I created this diagram to summarize and illustrate the serverless microservices patterns we will be discussing in this book:

- **Communication styles**: How services communicate together and externally.
- **Decomposition pattern**: Creating a service that is loosely coupled by business capability or bounded context.
- **Data management**: Deals with local and shared data stores.
- **Queries and messaging**: Looks at events and messages that are sent between microservices, and how services are queried efficiently.
- **Deployment**: Where ideally we would like uniform and independent deployments, you also don't want developers to re-create a new pipeline for each bounded context or microservice.
- **Observability and discovery**: Being able to understand whether a service is functioning correctly, monitor and log activity allow you to drill down if there are issues. You also want to know and monitor what is currently running for cost and maintenance reasons, for example.
- **Security**: This is critical for compliance, data integrity, data availability, and potential financial damage. It's important to have different encryption, authentication, and authorization processes in place.

Next we will have a look at the communication styles and decomposition pattern first.

Communication styles and decomposition microservice patterns

In this section, we will discuss two microservice patterns, called **communication styles and decomposition**, with a sufficient level of detail that you will be able to discuss them with other developers, architects, and DevOps.

Communication styles

Microservice applications are distributed by nature, so they heavily rely on the authorizations network. This makes it important to understand the different communications styles available. These can be to communicate with each other but also with the outside world. Here are some examples:

- **Remote procedure calls**: It used to be popular for Java to use **Remote Method Invocation (RMI)**, which is a tight coupling between client and servers with a non-standard protocol, which is one limitation. In addition, the network is not reliable and so traditional RMIs should be avoided. Others, such as the SOAP interface, and a client generated from the **Web Service Definition Language (WSDL)**, are better but are seen as heavy weight, compared to **REpresentational State Transfer (REST)** APIs that have widely been adopted in microservices.
- **Synchronous communication**: It is simpler to understand and implement; you make a request and get a response. However, while waiting for the response, you may also be blocking a connection slot and resources, limiting calls from other services:

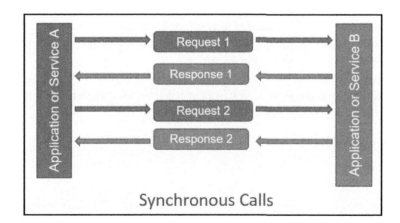

- **Asynchronous communication**: With asynchronous communication, you make the request and then get the response later and sometimes out of order. These can be implemented using callbacks, `async/await`, or `promise` in Node.js or Python. However, there are many design considerations in using `async`, especially if there are failures that need monitoring. Unlike most synchronous calls, these are non-blocking:

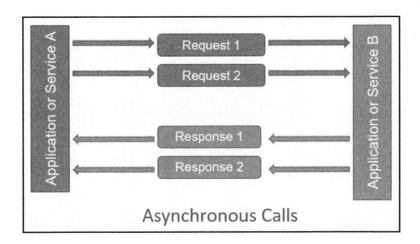

When dealing with communications, you also need to think about whether your call is blocking or non-blocking. For example, writing metrics from web clients to a NoSQL database using blocking calls could slow down your website.

You need to think about dealing with receiving too many requests and throttling them to not overwhelm your service, and look at failures such as retires, delays, and errors.

When using Lambda functions, you benefit from AWS-built event source and spinning up a Lambda per request or with a micro-batch of data. In most cases, synchronous code is sufficient even at scale, but it's important to understand the architecture and communication between services when designing a system, as it is limited by bandwidth, and network connections can fail.

One-to-one communication microservice patterns

At an individual microservice level, the data management pattern is composed of a suite of small services, with its own local data store, communicating with a REST API or via publish/subscribe:

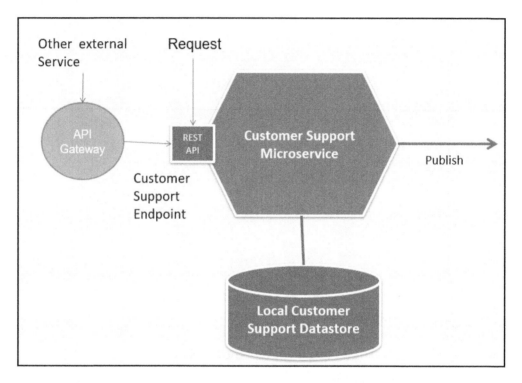

API Gateway is a single entry point for all clients, and tailored for them, allowing changes to be decoupled from the main microservice API, which is especially useful for external-facing services.

One-to-one request/response can be sync or async. If they are sync, they can have a response for each request. If the communication is async, they can have an async response or async notification. Async is generally preferred and much more scalable, as it does not hold an open connection (non-blocking), and makes better use of the **central processing unit (CPU)** and **input/output (I/O)** operations.

We will go into further detail on the data-management patterns later in the book, where we will be looking at how microservices integrate in a wider ecosystem.

Many-to-many communication microservice patterns

For many-to-many communication, we use publish/subscribe, which is a messaging pattern. This is where senders of messages, called publishers, do not program the messages to be sent directly to specific receivers; rather, the receiver needs to subscribe to the messages. It's a highly scalable pattern as the two are decoupled:

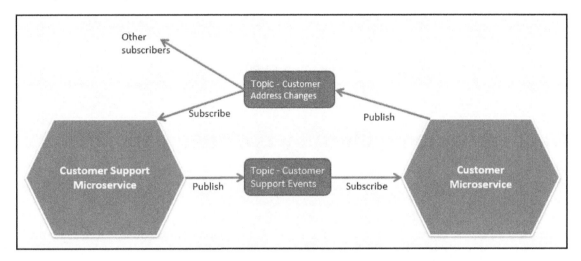

Asynchronous messaging allows a service to consume and act upon the events, and is a very scalable pattern as you have decoupled two services: the publisher and the subscriber.

Decomposition pattern by business capability

How do you create and design microservices? If you are migrating existing systems, you might look at decomposing a monolith or application into microservices. Even for new a green-field project, you will want to think about the microservices that are required:

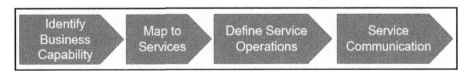

First, you identify the business capability, that is, *what* an organization does in order to generate value, rather than *how*. That is, you need to analyze purpose, structure, and business processes. Once you identify the business capabilities, you define a service for each capability or capability group. You then need to add more details to understand what the service does by defining the available methods or operations. Finally, you need to architect how the services will communicate.

The benefit of this approach is that it is relatively stable as it is linked to what your business offers. In addition, it is linked to processes and stature.

The drawbacks are that the data can span multiple services, it might not be optimum communication or shared code, and needs a centralized enterprise-language model.

Decomposition pattern by bounded context

There are three steps to apply the decomposition pattern by bounded context: first, identify the domain, which is *what* an organization does. Then identify the subdomain, which is to split intertwined models into logically-separated subdomains according to their actual functionality. Finally, find the bounded context to mark off where the meaning of every term used by the domain model is well understood. Bounded context does not necessarily fall within only a single subdomain. The three steps are as follows:

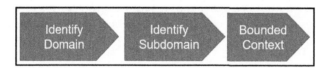

The benefits of this pattern are as follows:

- Use of Ubiquitous Language where you work with domain experts, which helps with wider communication.
- Teams own, deploy, and maintain services, giving them flexibility and a deeper understanding within their bounded context. This is good because services within it are most likely to talk to each other.
- The domain is understood by the team with a representative domain expert. There is an interface that abstracts away of a lot of the implementation details for other teams.

There are a few drawbacks as well:

- It needs domain expertise.
- It is iterative and needs to be **continuous integration** (CI) to be in place.

- Overly complex for a simple domain, dependent on Ubiquitous Language and domain expert.
- If a polyglot approach was used, it's possible no one knows the tech stack any more. Luckily, microservices should be smaller and simpler, so these can be rewritten.

More details can be found in the following books:

- *Building-microservices*, Sam Newman (2015)
- *Domain-Driven Design: Tackling Complexity in the Heart of Software*, Eric Evans (2003)
- *Implementing Domain-Driven Design*, Vaughn Vernon (2013)

Serverless computing in AWS

Serverless computing in AWS allows you to quickly deploy event-driven computing in the cloud. With serverless computing, there are still servers but you don't have the manage them. AWS automatically manages all the computing resources for you, as well as any trigger mechanisms. For example, when an object gets written to a bucket, that would trigger an event. If another service writes a new record to an Amazon DynamoDB table, that could trigger an event or an endpoint to be called.

The main idea of using event-driven computing is that it easily allows you to transform data as it arrives into the cloud, or we can perform data-driven auditing analysis notifications, transformations, or parse **Internet of Things (IoT)** device events. Serverless also means that you don't need to have an always-on running service in order to do that, you can actually trigger it based on the event.

Overview of some of the key serverless services in AWS

Some key serverless services in AWS are explained in the following list:

- **Amazon Simple Storage Service (S3)**: A distributed web-scale object store that is highly scalable, highly secure, and reliable. You only pay for the storage that you actually consume, which makes it beneficial in terms of pricing. It also supports encryption, where you can provide your own key or you can use a server-side encryption key provided by AWS.

- **Amazon DynamoDB**: A fully-managed NoSQL store database service that is managed by AWS and allows you to focus on writing the data out to the data store. It's highly durable and available. It has been used in gaming and other high-performance activities, which require low latency. It uses SSD storage under the hood and also provides partitioning for high availability.
- **Amazon Simple Notification Service (SNS)**: A push-notification service that allows you to send notifications to other subscribers. These subscribers could be email addresses, SNS messages, or other queues. The messages would get pushed to any subscriber to the SNS service.
- **Amazon Simple Queue Service (SQS)**: A fully-managed and scalable distributed message queue that is highly available and durable. SQS queues are often subscribed to SNS topics to implement the distributed publish-subscribe pattern. You pay for what you use based on the number of requests.
- **AWS Lambda**: The main idea is you write your business logic code and it gets triggered based on the event sources you configure. The beauty is that you only pay for when the code is actually executed, down to the 100 milliseconds. It automatically scales and is highly available. It is one of the key components to the AWS serverless ecosystem.
- **Amazon API Gateway**: A managed API service that allows you to build, publish, and manage APIs. It performs at scale and allows you to also perform caching, traffic throttling, and caching in edge locations, which means they're localized based on where the user is located, minimizing overall latency. In addition, it integrates natively with AWS Lambda functions, allowing you to focus on the core business logic code to parse that request or data.
- **AWS Identity and Access Management (IAM)**: The central component of all security is IAM roles and policies, which are basically a mechanism that's managed by AWS for centralizing security and federating it to other services. For example, you can restrict a Lambda to only read a specific DynamoDB table, but not have the ability to write to the same DynamoDB table or deny read/write access any other tables.
- **Amazon CloudWatch**: A central system for monitoring services. You can, for example, monitor the utilization of various resources, record custom metrics, and host application logs. It is also very useful for creating rules that trigger a notification when specific events or exceptions occur.

- **AWS X-Ray**: A service that allows you to trace service requests and analyze latency and traces from various sources. It also generates service maps, so you can see the dependency and where the most time is spent in a request, and do root cause analysis of performance issues and errors.

- **Amazon Kinesis Streams**: A steaming service that allows you to capture millions of events per second that you can analyze further downstream. The main idea is you would have, for example, thousands of IoT devices writing directly to Kinesis Streams, capturing that data in one pipe, and then analyzing it with different consumers. If the number of events goes up and you need more capacity, you can simply add more shards, each with a capacity of 1,000 writes per second. It's simple to add more shards as there is no downtime, and they don't interrupt the event capture.

- **Amazon Kinesis Firehose**: A system that allows you to persist and load streaming data. It allows you to write to an endpoint that would buffer up the events in memory for up to 15 minutes, and then write it into S3. It supports massive volumes of data and also integrates with Amazon Redshift, which is a data warehouse in the cloud. It also integrates with the Elasticsearch service, which allows you to query free text, web logs, and other unstructured data.

- **Amazon Kinesis Analytics**: Allows you to analyze data that is in Kinesis Streams using **structured query language (SQL)**. It also has the ability to discover the data schema so that you can use SQL statements on the stream. For example, if you're capturing web analytics data, you could count the daily page view data and aggregate them up by specific `pageId`.

- **Amazon Athena**: A service that allows you to directly query S3 using a schema on read. It relies on the AWS Glue Data Catalog to store the table schemas. You can create a table and then query the data straight off S3, there's no spin-up time, it's serverless, and allows you to explore massive datasets in a very flexible and cost-effective manner.

Among all these services, AWS Lambda is the most widely used serverless service in AWS. We will discuss more about that in the next section.

AWS Lambda

The key serverless component in AWS is called **AWS Lambda**. A Lambda is basically some business logic code that can be triggered by an event source:

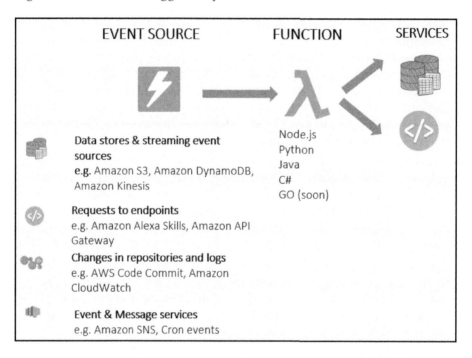

A **data event source** could be the put or get of an object to an S3 bucket. **Streaming event sources** could be new records that have been to a DynamoDB table that trigger a Lambda function. Other streaming event sources include Kinesis Streams and SQS.

One example of **requests to endpoints** are Alexa skills, from Alexa echo devices. Another popular one is Amazon API Gateway, when you call an endpoint that would invoke a Lambda function. In addition, you can use changes in AWS CodeCommit or Amazon Cloud Watch.

Finally, you can trigger different **events and messages based** on SNS or different cron events. These would be regular events or they could be notification events.

The main idea is that the integration between the event source and the Lambda is managed fully by AWS, so all you need to do is write the business logic code, which is the function itself. Once you have the function running, you can run either a transformation or some business logic code to actually write to other services on the right of the diagram. These could be data stores or invoke other endpoints.

In the serverless world, you can implement sync/asyc requests, messaging or event stream processing much more easily using AWS Lambdas. This includes the microservice communication style and data-management patterns we just talked about.

Lambda has two types of event sources types, non-stream event sources and stream event sources:

- **Non-stream event sources**: Lambdas can be invoked asynchronously or synchronously. For example, SNS/S3 are asynchronous but API Gateway is sync. For sync invocations, the client is responsible for retries, but for async it will retry many times before sending it to a **Dead Letter Queue** (**DLQ**) if configured. It's great to have this retry logic and integration built in and supported by AWS event sources, as it means less code and a simpler architecture:

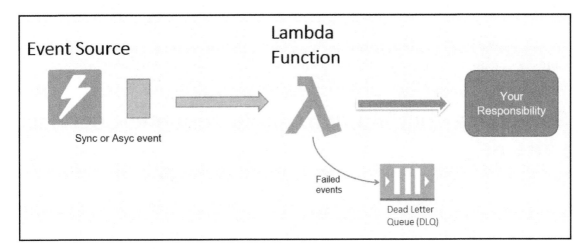

- **Stream event sources**: The Lambda is invoked with micro-batches of data. In terms of concurrency, there is one Lambda invoked in parallel per shard for Kinesis Streams or one Lambda per partition for DynamoDB Stream. Within the lambda, you just need to iterate over the Kinesis Streams, DynamoDB, or SQS data passed in as JSON records. In addition, you benefit from the AWS built-in streams integration where the Lambda will poll the stream and retrieve the data in order, and will retry upon failure until the data expires, which can be up to seven days for Kinesis Streams. It's also great to have that retry logic built in without having to write a line of code. It is much more effort if you had to build it as a fleet of EC2 or containers using the AWS Consumer or Kinesis SDK yourself:

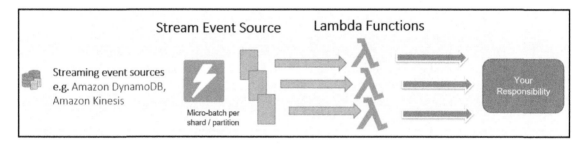

In essence, AWS is responsible for the invocation and passing in the event data to the Lambda, you are responsible for the processing and the response of the Lambda.

Serverless computing to implement microservice patterns

Here is an overview diagram of some of the serverless and managed services available on AWS:

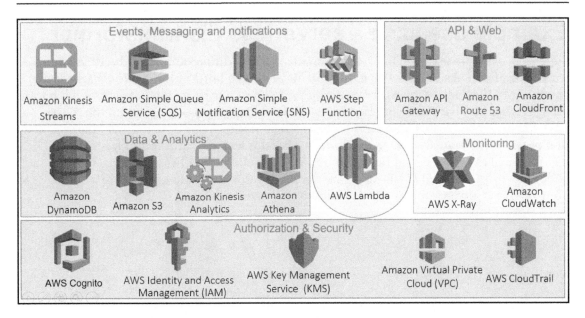

Leveraging AWS-managed services does mean additional vendor lock-in but helps you reduce non business differentiating support and maintenance costs. But also to deploy your applications faster as the infrastructure can be provisioned or destroyed in minutes. In some cases, when using AWS-managed services to implement microservices patterns, there is no need for much code, only configuration.

We have services for the following:

- **Events, messaging, and notifications**: For async publish/subscribe and coordinating components
- **API and web**: To create APIs for your serverless microservices and expose it to the web
- **Data and analytics**: To store, share, and analyze your data
- **Monitoring**: Making sure your microservices and stack are operating correctly
- **Authorization and security**: To ensure that your services and data is secure, and only accessed by those authorized

At the center is AWS Lambda, the glue for connecting services, but also one of the key places for you to deploy your business logic source code.

Example use case – serverless file transformer

Here is an example use case, to give you an idea of how different managed AWS systems can fit together as a solution. The requirements are that a third-party vendor is sending us a small 10 MB file daily at random times, and we need to transform the data and write it to a NoSQL database so it can be queried in real time. If there are any issues with the third-party data, we want to send an alert within a few minutes. Your boss tells you that you they don't want to have an always-on machine just for this task, the third party has no API development experience, and there is a limited budget. The head of security also finds out about this project and adds another constraint. They don't want to give third-party access to your AWS account beyond one locked-down S3 bucket:

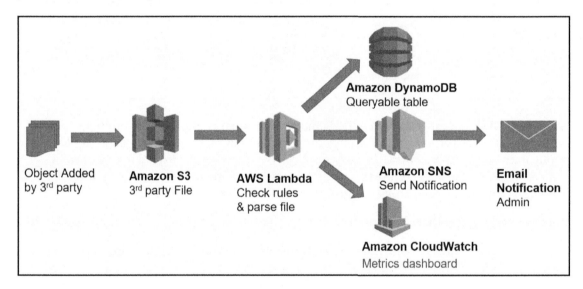

This can be implemented as an event-driven serverless stack. On the left, we have an S3 bucket where the third party has access to drop their file. When a new object is created, that triggers a Lambda invocation via the built-in event source mapping. The Lambda executes code to transform the data, for example, extracts key records such as user_id, date, and event type from the object, and writes them to a DynamoDB table. The Lambda sends summary custom metrics of the transformation, such as number of records transformed and written to CloudWatch metrics. In addition, if there are transformation errors, the Lambda sends an SNS notification with a summary of the transformation issues, which could generate an email to the administrator and third-party provider for them to investigate the issue.

Setting up your serverless environment

If you already have an AWS account and configured it locally you can skip this section, but for security reasons, I recommend you enable **Multi-Factor Authentication (MFA)** for console access and do not use the root user account keys for the course.

There are three ways to access resources in AWS:

- AWS Management Console is a web-based interface to manage your services and billing.
- AWS Command Line Interface is a unified tool to manage and automate all your AWS services.
- The software-development kit in Python, JavaScript, Java, .NET, and GO, which allows you to programmatically interact with AWS.

Setting up your AWS account

It's very simple to set up an account; all you need is about five minutes, a smartphone, and a credit card:

1. Create an account. AWS accounts include 12 months of Free Tier access: `https://aws.amazon.com/free/`.
2. Enter your name and address.
3. Provide a payment method.
4. Verify your phone number.

This will create a root account, I recommend you only use it for billing and not development

Setting up MFA

I recommend you use MFA as it adds an extra layer of protection on top of your username and password. It's free using your mobile phone as a Virtual MFA Device (`https://aws.amazon.com/iam/details/mfa/`). Perform the following steps to set it up:

1. Sign into the AWS Management Console: `https://console.aws.amazon.com`.
2. Choose **Dashboard** on the left menu.
3. Under **Security Status**, expand **Activate MFA on your root account**.
4. Choose **Activate MFA** or **Manage MFA.**

5. In the wizard, choose **Virtual MFA device**, and then choose **Continue.**
6. Install an MFA app such as Authy (`https://authy.com/`).
7. Choose **Show QR code** then scan the QR code with you smartphone. Click on the account and generate an Amazon six-digit token.
8. Type the six-digit token in the **MFA code 1** box.
9. Wait for your phone to generate a new token, which is generated every 30 seconds.
10. Type the six-digit token into the **MFA code 2** box.
11. Choose **Assign MFA:**

Setting up a new user with keys

For security reasons, I recommend you use the root account only for billing! So, the first thing is to create another user with fewer privileges:

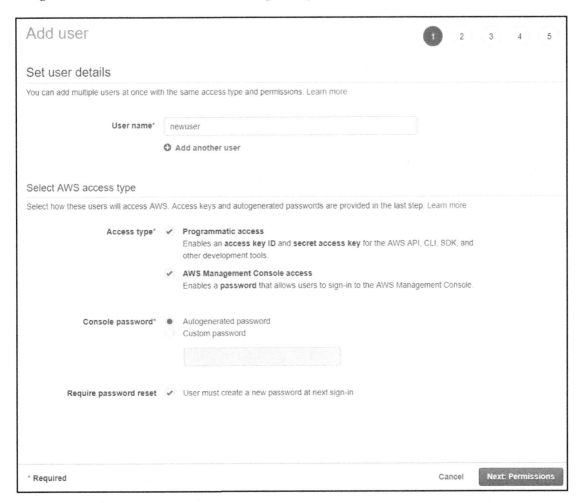

Create a user with the following steps:

1. Sign into the AWS Management console (https://console.aws.amazon.com/).
2. Choose **Security, Identity, & Compliance > IAM** or search for **IAM** under **Find services.**
3. In the IAM page, choose **Add User.**
4. For **User name**, type new user on the **set user details** pane.

5. For **Select AWS access Type**, select the check boxes next to **Programmatic access, AWS Console access**. Optionally select **Autogenerated password** and **Require password rest.**

6. Choose **Next: Permissions:**

Follow these steps to set the permission for the new user:

1. Choose **Create group.**

2. In the **Create group** dialog box, type `Administrator` for new group name.

3. In policy list, select the checkbox next to **AdministratorAccess** (note that, for non-proof of concept or non-development AWS environments, I recommend using more restricted access policies).

4. Select **Create group.**

5. Choose refresh and ensure the checkbox next to **Administrator** is selected.
6. Choose **Next: Tags.**
7. Choose **Next: Review.**
8. Choose **Create user.**
9. Choose **Download .csv** and take a note of the keys and password. You will need these to access the account programmatically and log on as this user.
10. Choose **Close.**

 More details on creating a new user can be found at `https://docs.aws.` `amazon.com/IAM/latest/UserGuide/getting-started_create-admin-` `group.html`.

As with the root account, I recommend you enable MFA:

1. In the Management Console, choose **IAM** | **User** and choose the **newuser.**
2. Choose the **Security Credentials** tab, then choose **Manage** next to **Assigned MFA device Not assigned**.
3. Choose a virtual MFA device and choose **Continue**.
4. Install an MFA application such as Authy (`https://authy.com/`).
5. Choose **Show QR code** then scan the QR code with you smartphone. Click on the Account and generate an Amazon six-digit token.
6. Type the six-digit token in the **MFA code 1** box.
7. Wait for your phone to generate a new token, which is generated every 30 seconds.
8. Type the six-digit token into the **MFA code 2** box.
9. Choose **Assign MFA**.

Managing your infrastructure with code

A lot can be done with the web interface in the AWS Management Console. It's a good place to start and help you to understand what you are building, but most often it is not recommended for production deployments as it is time-consuming and prone to human error. Best practice is to deploy and manage your infrastructure using code and configuration only. We will be using the AWS **Command-line Interface (CLI)**, bash shell scripts, and Python 3 throughout this book, so let's set these up now.

Installing bash on Windows 10

Please skip this step if you are not using Windows.

Using bash (Unix shell) makes your life much easier when deploying and managing your serverless stack. I think all analysts, data scientists, architects, administrators, database administrators, developers, DevOps, and technical people should know some basic bash and be able to run shell scripts, which are typically used on Linux and Unix (including the macOS Terminal).

Alternatively, you can adapt the scripts to use MS-DOS or PowerShell, but it's not something I recommended, given that bash can now run natively on Windows 10 as an application, and there are many more examples online in bash.

Note that I have stripped off the \r or carriage returns, as they are illegal in shell scripts. You can use something such as Notepad++ (https://notepad-plus-plus.org/) on Windows if you want to view the carriage returns in your files properly. If you use traditional Windows Notepad, the new lines may not be rendered at all, so use Notepad++, Sublime (https://www.sublimetext.com/), Atom (https://atom.io/), or another editor.

A detailed guide on how to install Linux Bash shell on Windows 10 can be found at https://www.howtogeek.com/249966/how-to-install-and-use-the-linux-bash-shell-on-windows-10/. The main steps are as follows:

1. Navigate to **Control Panel** | **Programs** | **Turn Windows Features On Or Off**.
2. Choose the check box next to the **Windows Subsystem for Linux** option in the list, and then Choose **OK**.
3. Navigate to **Microsoft Store** | **Run Linux on Windows** and select **Ubuntu**.
4. Launch Ubuntu and set up a root account with a username and password the Windows C:\ and other drives are already mounted, and you can access them with the following command in the Terminal:

```
$ cd /mnt/c/
```

Well done, you now have full access to Linux on Windows!

Updating Ubuntu, installing Git and Python 3

Git will be used later on in this book:

```
$ sudo apt-get update
$ sudo apt-get -y upgrade
$ apt-get install git-core
```

The Lambda code is written in Python 3.6. `pip` is a tool for installing and managing Python packages. Other popular Python package and dependency managers are available, such as Conda (`https://conda.io/docs/index.html`) or Pipenv (`https://pipenv.readthedocs.io/en/latest/`), but we will be using pip as it is the recommended tool for installing packages from the Python Package Index PyPI (`https://pypi.org/`) and is the most widely supported:

```
$ sudo apt -y install python3.6
$ sudo apt -y install python3-pip
```

Check the Python version:

```
$ python --version
```

You should get Python version 3.6+.

The dependent packages required for running, testing, and deploying the severless microservices are listed in `requirements.txt` under each project folder, and can be installed using `pip`:

```
$ sudo pip install -r /path/to/requirements.txt
```

This will install the dependent libraries for local development, such as Boto3, which is the Python AWS **Software Development Kit (SDK)**.

In some projects, there is a file called `lambda-requirements.txt`, which contains the third-party packages that are required by the Lambda when it is deployed. We have created this other `requirements` file as the Boto3 package is already included when the Lambda is deployed to AWS, and the deployed Lambda does not need testing-related libraries, such as `nose` or `locust`, which increase the package size.

Installing and setting up the AWS CLI

The AWS CLI is used to package and deploy your Lambda functions, as well as to set up the infrastructure and security in a repeatable way:

```
$ sudo pip install awscli --upgrade
```

You created a user called `newuser` earlier and have a `crednetials.csv` file with the AWS keys. Enter them by running `aws configure`:

```
$ aws configure
AWS Access Key ID: <the Access key ID from the csv>
AWS Secret Access Key: <the Secret access key from the csv>
Default region name: <your AWS region such as eu-west-1>
Default output format: <optional>
```

More details on setting up the AWS CLI are available in the AWS docs (`https://docs.aws.amazon.com/lambda/latest/dg/welcome.html`).

To choose your AWS Region, refer to AWS Regions and Endpoints (`https://docs.aws.amazon.com/general/latest/gr/rande.html`). Generally, those in the USA use `us-east-1` and those in Europe use `eu-west-1`.

Summary

In this chapter, we got an overview of monolithic and microservices architectures. We then talked about the design patterns and principles and how they relate to serverless microservices. We also saw how to set up the AWS and development environment that will be used in this book.

In the next chapter, we will create a serverless microservice that exposes a REST API and is capable of querying a NoSQL store built using API Gateway, Lambda, and DynamoDB.

2
Creating Your First Serverless Data API

In this chapter, we will build a complete serverless microservice, accessible via a REST API, and capable of querying a NoSQL database. We will start by discussing and creating the **Amazon Web Services** (**AWS**) security infrastructure to ensure restricted access to AWS resources. We will then create, add records to, and query a NoSQL database, first using the Management Console, then using Python. Then, we will go over the code used in the Lambda function in Python and API Gateway integration. Finally, we will deploy it and test that the API is working.

The following topics will be covered in this chapter:

- Overview of security in AWS
- Securing your serverless microservice
- Building a serverless microservice data API
- Setting up Lambda security in the AWS management console
- Creating and writing to a NoSQL database called DynamoDB using AWS
- Creating and writing to a NoSQL database called DynamoDB using Python
- Creating a Lambda to query DynamoDB
- Setting up API Gateway and integrating it with a Lambda Proxy
- Connecting API Gateway, Lambda, and DynamoDB
- Cleaning-up

Overview of security in AWS

We will start with a discussion on security and how to set it up correctly in AWS.

Why is security important?

You might have heard of ransomware, cyber attacks, or security breaches recently and you would not want your organization to be affected by these. Some of these are shown as follows:

```
Insecure Communication                                      Litigation Costs

   Misconfiguration                Social engineering            Data Loss

   Unpatched Vulnerabilities              Viruses            Reputation Costs
                              Hacking
   Insecure Disposal                                           Data Breaches

                                  Malware              Financial Costs
                                         Hacktivist
   Missing Updates              Ransomware

   Insecure Storage         Phishing   Rootkit        Business Disruption

   Leaked Keys                  Cracking             Report Incidents

   Weak Physical Security   Distributed Denial of Service Attack (DDoS)   Ransom Payments
```

Systems not being configured correctly, missing updates, or using insecure communication can lead to them being hacked or being subject to a ransomware demand. This can result in litigation costs, data loss or leaks, and financial costs to your organization.

There are many reasons for ensuring your systems are secure, including the following:

- **Compliance**: Compliance with the law, regulations, and standards, for example, the EU **General Data Protection Regulation (GDPR)**, the **Health Information Portability and Accountability Act (HIPAA)**, and the Federal Trade Commission Act.

- **Data integrity**: If systems aren't secure, data could be stripped or tampered with, meaning you can no longer trust the customer data or financial reporting.

- **Personally Identifiable Information (PII)**: Consumers and clients are aware of your privacy policy. Data should be securely protected, anonymized, and deleted when no longer required.

- **Data availability**: Data is available to authorized users, but if, for example, a natural disaster occurred in your data center, what would happen in terms of accessing data?

A lot of security in AWS stems from configuration and having the correct architecture, so it's important to understand the following subset of important security related terms:

- **Security in transit:** For example, HTTPS SSL—think of it as the padlock on your browser
- **Security at rest:** For example, data encryption, where only a user with a key can read the data in a data store
- **Authentication**: For example, a process to confirm the user or system are who they are meant to be
- **Authorization:** For example, permissions and control mechanisms to access specific resources

Security by design principles

There are many security standards, principles, certifications, and guidance—probably enough to fill a few books. Here is one that I found practical and useful, from the **Open Web Application Security Project (OWASP)** at `https://www.owasp.org`. The OWASP security by design principles (`https://www.owasp.org/index.php/Security_by_Design_Principles`) apply to any system, application, or service, helping to make them more secure by design, and that includes serverless computing. Even if there are no servers that need managing with serverless, you still need to ensure your architecture, integration, configuration, and code adhere to the following principles:

- **Minimize attack surface area**: Every added feature is a risk—ensure they are secure, for example, delete any Lambdas that are no longer being used.
- **Establish secure defaults**: These have defaults for every user, Identity and Access Management policy, and serverless stack component.
- **Principle of least privilege**: The account or service has the least amount of privilege required to perform its business processes, for example, if a Lambda only needs read access to a table, then it should have no more access than that.
- **Principle of defense in depth**: Have different validation layers and centralized audit controls.
- **Fail securely**: This ensures that if a request or transformation fails, it is still secure.
- **Don't trust services**: Especially third parties, external services, or libraries, for example, JavaScipt and Node.js libraries infected with malware.
- **Separation of duties**: Use a different role for a different task, for example, administrators should not be users or system users.

- **Avoid security by obscurity**: This is generally a bad idea and a weak security control. Instead of relying on the architecture or source code being secret, instead rely on other factors, such as good architecture, limiting requests, and audit controls.
- **Keep security simple**: Don't over-engineer; use simple architectures and design patterns.
- **Fix security issues correctly**: Fix issues promptly and add new tests.

Keep these principles in mind when building any serverless microservices.

AWS Identity and Access Management

Identity and Access Management (IAM), is a central location where you can manage users' security credentials, such as passwords, access keys, and permission policies, that control access to the AWS services and resources. We are going to talk about the most relevant IAM resources—policies, users, groups, and roles—but first, we will talk about the JSON (https://www.json.org/) format as it is used in IAM policies.

JavaScript object notation

JSON, or JavaScript object notation, is a standard data format that is used in REST APIs and microservices. It can be read by humans but also by machines. So, humans can actually understand the values and, also, machines can automatically parse the data. The data objects consist of attribute-value pairs and array data types. The data type values supported are number, string, Boolean, array, object, and null, as shown in the following code:

```
{
  "firstName": "John",
  "lastName": "Smith",
  "age": 27,
  "address": {
    "city": "New York",
    "postalCode": "10021"
  },
  "phoneNumbers": [
    {
      "type": "home",
      "number": "212 555-1234"
    },
    {
      "type": "mobile",
      "number": "123 456-7890"
```

```
        }
    ]
}
```

The preceding code is an example of details related to John Smith. You can see the first name is the key and the string value is John. In that way, you have two strings separated by a colon. So, you see that John Smith is 27 years old and the city he lives in is New York, with the postal code 10021, and you can see that he has two phone numbers. One is his home phone number and the other one is a mobile phone number. JSON is very descriptive and can easily be parsed programmatically.

You can see that it doesn't have to be flat either. It can also be hierarchical and has the keys built into the data. You can also very easily add new phone numbers to JSON data and extend the schema, without breaking the model, unlike other formats such as **comma-separated variable (CSV)** files. JSON is supported natively in the standard Python JSON library, but you can also use other libraries. What I like about it is that there is native mapping to Python data types.

IAM policies

IAM policies are JSON documents that define effects, actions, resources, and conditions, as shown in the following code:

```
{
    "Version": "2012-10-17",
    "Statement": {
        "Effect": "Allow",
        "Action": [
                "dynamodb:GetItem",
                "dynamodb:Scan",
                "dynamodb:Query"],
        "Resource": "arn:aws:dynamodb:eu-west-
                1:123456789012:table/Books",
        "Condition": {
            "IpAddress": {
                "aws: SourceIp": "10.70.112.23/16"
            }
        }
    }
}
```

This is an example of a JSON document that will grant read access to a DynamoDB table called `Books` only if the request originates from a specific **Classless Inter-Domain Routing (CIDR)**, `10.70.112.23/16`, that is, in **Internet Protocol address (IP address)** version 4 ranges from `10.70.0.0` to `10.70.255.255`.

There is also a visual editor that allows you to create these, or you can do so manually by editing the actual JSON document itself. For example, we created a new user earlier in the book and gave them administrator rights using an AWS managed policy, but you can create your own, as shown. My recommendation is to use AWS managed policies when possible, unless it's a resource for which access can, or should, be more restricted, such as a DynamoDB table or an S3 bucket.

IAM users

IAM users are people or services that interact with AWS. We actually set up a new user in `Chapter 1`, *Serverless Microservices Architectures and Patterns*. They can have access to the AWS Management Console via a password with multi-factor authentication, and/or they may have an access key for programmatic access using the command-line interface or the AWS **software development kits (SDKs)**. You can attach one or more IAM policies directly to a user to grant them access to a resource or service. A policy could be like what we have just shown you, for granting read access to a DynamoDB table called `Books` from a specific originating IP range.

IAM groups

IAM groups are used to mimic this security feature in your organization groups. You could think of it as active directory groups. For example, in your organization, you would have administrators, developers, and testers. To create a group, you can use the AWS Management Console, the SDK, or the CLI. Once you have created a group, you can attach it to a user or, alternatively, you can create one when you are creating a new user. I tend to attach IAM policies to a group, and then assign groups to users, as it makes it much easier to manage and standardizes access. For example, I can assign the data science group to new joiners of the team knowing that it's identical to the other users. Equally, if someone leaves, then their magical policies will not be deleted with them!

IAM roles

IAM roles are similar to IAM users, in that they can have a policy attached to them but they can be assumed by anyone who needs access in a so-called trusted entity. In that way, you can delegate access to users, applications, or services, without having to give them a new AWS key, as they use temporary security tokens through this trusted entity. For example, without actually having to share any keys, and purely using the roles, you could grant a third party read access to an S3 bucket only and nothing else within your AWS environment.

Securing your serverless microservices

In this section, we will discuss security in detail.

Lambda security

As we discussed earlier, AWS Lambda is the central component in a serverless stack, or the integration connector with your custom code, triggered by events between AWS managed services. A Lambda function always has an execution IAM role associated with it, and using policies attached to that role is one of the best, and most secure, ways to deny or grant it access to other AWS resources. The beauty is that there is no need to manage or exchange any keys or passwords for a lot of the AWS managed services, such as S3, DynamoDB, and Kinesis Stream. There are some exceptions, such as some of the Amazon **Relational Database Service** (**RDS**), such as SQL Server, but MySQL or PostgreSQL do support IAM database authentication. The following diagram shows the workings of Lambda functions:

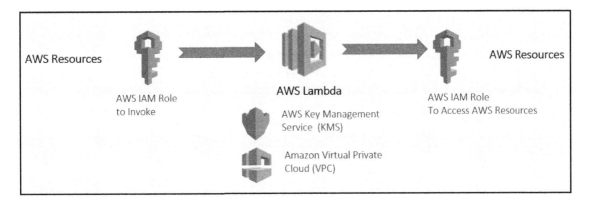

As the preceding diagram shows, there are generally two IAM roles involved in a Lambda function:

- Invoking the Lambda, for example, from API Gateway or AWS Step Functions
- Granting read and write access to AWS resources, for example, granting a Lambda read access to to a DynamoDB table

In addition, note the following:

- **Key Management Service** (**KMS**) can be used for the encryption/decryption of data at rest in DynamoDB or RDS, but also to encrypt passwords or keys, for example, should you need them to integrate with a third party API or database.
- Lambda is launched in a secure **Virtual Private Cloud** (**VPC**) by default. However, you can also run it inside your own private VPC if there are resources you need to access, such as ElastiCache clusters or RDS. You may also do so to add another layer of security.

API Gateway security

Let's have a look at the following diagram:

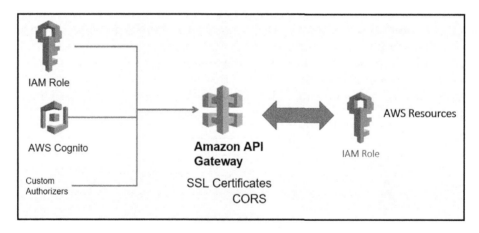

API Gateway can be used to create a public-facing API without authentication, but sometimes you will want to restrict access. The following are are three different ways of controlling who can call the request authorization API:

- IAM roles and policies can be used to grant access to an API, where API Gateway verifies the caller's signature on the requests transparently.

- An Amazon Cognito user pool controls who can access the API. The user or service will have to first sign in to access the API.
- API Gateway custom authorizers request, for example, a bearer token, and use a Lambda function to check whether the client is authorized to call the API.

Within API Gateway, note the following:

- If you get requests from a domain other than the API's own domain, you must enable **cross-origin resource sharing (CORS)**.
- Client-side SSL certificates are also supported, for example, to allow backend systems to validate that HTTP requests do originate from API Gateway and not another system.
- API Gateway may also need to be granted access via an IAM role, for example, should it need to write records to Kinesis Streams or invoke a Lambda function.
- Usage plans let you create API keys for customers, allowing you to limit and monitor usage. This could allow you to create a pay-per-usage API for your customers, for example.

DynamoDB security

Now let's look at the following diagram:

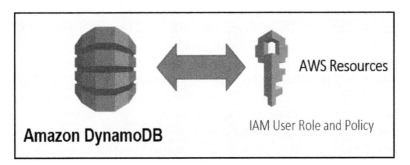

DynamoDB is an AWS-managed service and authorization is managed via an IAM permission policy. The IAM policy, which grants or deny access to DynamoDB, is attached to a specific IAM user or role, which can then access it. If you want to assume the role in one AWS account, we also have the option of delegating the permissions for the same, so that they can access a DynamoDB table in a different AWS account. The benefit in that scenario is that no keys are exchanged.

What I recommend is that you apply the **least privilege principle** when creating these policies for DynamoDB you lock them down as much as possible, which means avoiding the wildcard star for table access, such as using `"Resource":  "*"`. For example, in the policy documents, avoid giving read and write access to all the tables unless absolutely necessary. It is better to list specific actions, table names, and constraints explicitly when possible.

Monitoring and alerting

Now consider the following diagram:

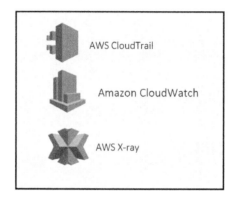

In general, it's important to monitor systems for any suspicious activity, or even spot any performance issues with systems. API Gateway, DynamoDB, and Lambda functions all have built-in support for CloudWatch and X-Ray for that specific task. CloudWatch allows you to track metrics and monitor log files, set specific alarms, and automatically react to changes in your AWS resources. X-Ray is a service that traces requests and can also generate specific service maps. The combination of these free systems gives you very good insight, an out-of-the-box view, into how your serverless system is performing. CloudTrail is another service that allows you to monitor all APIs and access to resources by any user or system.

Find out more

You will now have a much deeper understanding of security in AWS and why it's important for your organization.

If you want to find out more, here are some links to white papers and best practice guides. I recommend reading the following white papers:

- https://aws.amazon.com/whitepapers/aws-security-best-practices/
- https://aws.amazon.com/products/security/
- https://aws.amazon.com/whitepapers/#security
- http://docs.aws.amazon.com/IAM/latest/UserGuide/best-practices.html

Building a serverless microservice data API

In this section, we will look at the architecture and requirements of building a serverless microservice. The rest of the chapter is heavily hands-on in terms of configuration in the AWS Management Console, but also Python code. The Python code follows basic design patterns and is kept simple, so that you can understand and easily adapt it for your own use cases.

Serverless microservice data API requirements

We want to create a microservice that is able to serve a web visits count, for a specific event, of the total users browsing your website.

Query string

For a specific EventId passed into the request, we want to retrieve a daily web visits count (already collected by another system) as a response. We are going to be querying by EventId and startDate; where we want to retrieve all web visit counts after the startDate. The URL, with its parameter, will look as follows:

Here, you can see that we have EventId as resource 1234, and a startDate parameter formatted in the YYYYMMDD format. In this case, it's 20180102. That is the request.

We can either enter this request in the browser or programmatically, and the response that we want to get is from the NoSQL database, where the live data is stored. The actual response format will be in JSON coming back from the database, through a Lambda function, to be presented to either the user, or the other service that has queried the API programmatically. We also want this API to scale very easily and be very cost-effective; that is, we don't want to have a machine running all the time, for example, as that would cost money and need to be maintained. We only want to pay when actual requests are made.

Now the following diagram:

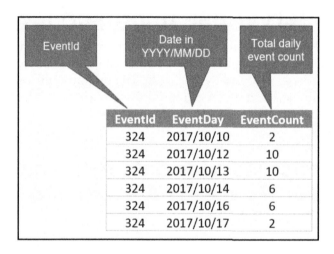

Here is an example of the time-series data that we're interested in, where we have EventId that is for event 324, and we have the date in the format EventDay, which is in October 2017, and we have a total EventCount of web events in the right-hand column in the table. You can see that on the October 10, 2017, EventId with 324 had an EventCount of 2, which means there was a total daily count of visits equal to 2 on that specific day for that event. The next day it's 0, as there is no entry for the 11th. Then, it increased to 10 on the 12th, 10 on the 13th, then it went down to 6, 0, 6, and 2. Note that when we have no data in the table for a specific date, then it's 0.

This is the data we want the API to provide as a response in JSON, so another client application can chart it, as shown in the following chart:

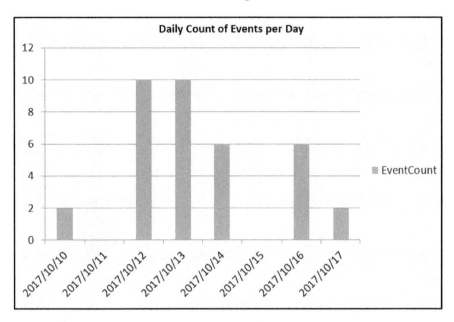

For example, if we plot the data in Excel, you would see this sort of chart, with the EventCount beginning at 2, with a gap on the October 11, where there was no visits for this specific user, then an increase to 10, 10, 6, then a gap on October 15, then up to 6 again on October 16.

Data API architecture

Now that we know the requirements and shape of the data that we want to return, we will talk about the overall architecture. Again, the whole stack will rely on the JSON format for data exchange between all the services.

The following diagram shows what we have in the the request:

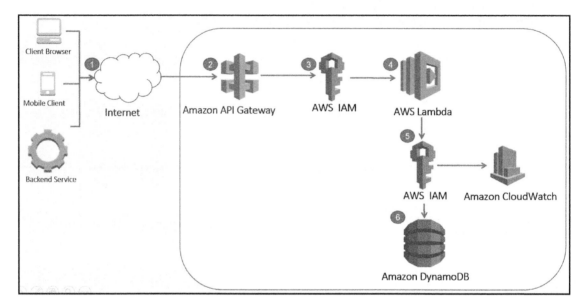

The request flow is as follows:

1. We have a client browser, mobile client, or a backend service on the internet.
2. That will query our API Gateway, passing the request with the EventId and the startDate as an optional URL parameter.
3. This will get authenticated through an AWS IAM role.
4. That will then launch the Lambda function.
5. The Lambda function will then use a role to access DynamoDB.
6. The Lambda will query Dynamo to search for that specific EventId. Optionally, the query will also include a specific startDate that is compared with EventDay. If EventDay is greater than startDate, then the records will be returned.

The following diagram shows what we have in the the response:

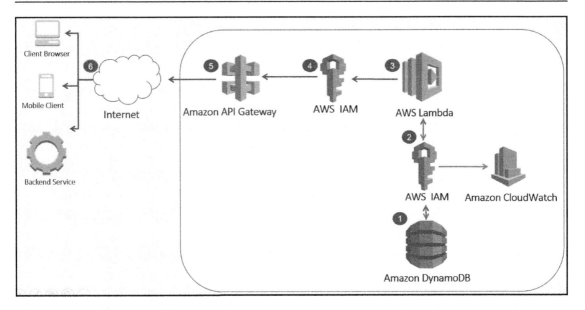

The response flow is as follows:

1. The data will be returned from DynamoDB, as shown at the bottom-right of the preceding diagram
2. This will be via the same IAM role associated with the Lambda function
3. The JSON records from DynamoDB are returned to the Lambda function, which parses it into a JSON response
4. That will get passed via the API Gateway role by the Lambda function
5. It is passed back into API Gateway
6. Eventually, it is returned to the client browser mobile client, or the backend service that made the initial request, so that it can be charted

We also have Amazon CloudWatch configured for monitoring requests, by providing dashboards for metrics and logs.

Setting up Lambda security in the AWS Management Console

We will be signing into the AWS Management Console. The reason we are using the Management Console first is to give you a better understanding of how Lambda functions work, and how they integrate with the other AWS services, such as API Gateway and DynamoDB. In later chapters, we will show you how to deploy Lambda functions using the AWS CLI. If you are a first-timer to Lambda, then I always find it useful to first create a full serverless stack manually in the Management Console to gain a better and deeper understanding than, say, have a magic command spin up the full AWS infrastructure!

We are going to first use the AWS Management Console to create the Lambda IAM role and policies, so that the Lambda function can access DynamoDB, and also write any logs or any statuses to CloudWatch. The Management Console, which we used earlier, in Chapter 1, *Serverless Microservices Architectures and Patterns*, allows you to centrally control all of the AWS services, create roles, and even create Lambda functions. In terms of the architecture for the serverless microservice, we are starting with the right part of the following diagram first, and building the rest step by step.

The following diagram shows the data API Lambda IAM:

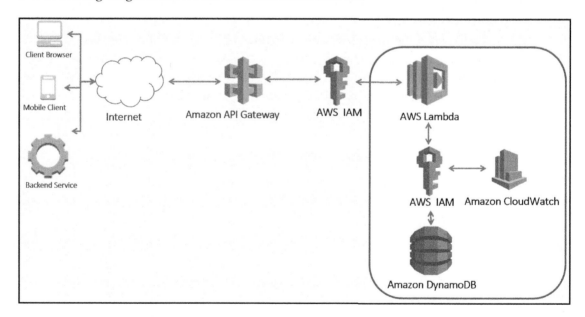

Create two IAM policies and attach them to a new Lambda IAM role next.

Creating an IAM policy

We are going to create the Lambda function and the IAM role and policies. The first thing that you need to do is to log on to the AWS Management Console. In IAM, we want to create the actual policy itself. You can click on **Create policies**, and we are going to use the JSON editor.

DynamoDB IAM policy

First, we need a policy allowing the Lambda function to read records from DynamoDB. We can do so by doing the following:

1. Sign into the AWS Management Console at `https://console.aws.amazon.com/`.
2. Choose **Security, Identity & Compliance | IAM**, or search for **IAM** under **Find services**.
3. In the IAM navigation pane, choose **Policies**.
4. Choose **Create policies**.
5. Choose the **JSON** tab.

> Rather than using the **JSON** view, you can also use, or switch to, the **Visual Editor** for creating a policy, but I prefer the JSON view, as the code can be source-controlled and deployed programmatically as we'll do later with the AWS CLI.

6. Type, or paste, the following JSON policy document:

```
{
    "Version": "2012-10-17",
    "Statement": [
        {
            "Effect": "Allow",
            "Action": [
                "dynamodb:BatchGetItem",
                "dynamodb:DescribeTable",
                "dynamodb:GetItem",
                "dynamodb:Query",
                "dynamodb:Scan"
            ],
            "Resource": [
                "arn:aws:dynamodb:<your-region>:<your-aws-
```

```
                               accountid>:table/user-visits"
                    ]
               }
          ]
     }
```

Update <your-region> with your AWS region, such as us-east-1, and update <your-aws-accountid> with your AWS account ID.

If you do not know your AWS account number, you can find it in the **Support Center** window, available from the top **Support** | **Support Center** menu in the AWS Management Console, as shown in the following screenshot:

7. Choose **Review Policy**.
8. On the **Review Policy** page, type dynamo-readonly-user-visits for the name.
9. Choose **Create Policy**.

This IAM policy, called **dynamo-readonly-user-visits**, will now be available under the **Filter policies** as **Customer managed**.

We talked about security being very important, and one way to ensure it is to apply the OWASP security by design principles, such as the principle of least privilege, as talked about earlier. Here, we do that by locking down the table access using a policy. You'll notice that I've restricted it to a specific name, dynamo table. For the policy name, it should be as descriptive and granular as possible to make it easier to maintain. I tend to have one policy per AWS resource where possible. I've used prefix dynamo-readonly so that it is obvious that you will only get the ability to read from one specific table, called user-visits.

Lambda IAM policy

Create a policy to be able to write logs and push metrics to CloudWatch:

1. Sign in to the AWS Management Console and open the IAM console at `https://console.aws.amazon.com/iam/`, if you're not signed in already.
2. In the IAM navigation pane, choose **Policies**.
3. Choose **Create policies**.
4. Choose the **JSON** tab.
5. Type or copy and paste the following JSON document:

```json
{
    "Version": "2012-10-17",
    "Statement": [
      {
        "Effect": "Allow",
        "Action": [
          "logs:CreateLogGroup",
          "logs:CreateLogStream",
          "logs:PutLogEvents",
          "logs:DescribeLogStreams"
        ],
        "Resource": [
          "arn:aws:logs:*:*:*"
        ]
      },
      {
        "Effect": "Allow",
        "Action": [
          "cloudwatch:PutMetricData"
        ],
        "Resource": "*"
      }
    ]
}
```

The main idea of this policy is to allow the Lambda function to create CloudWatch log groups and streams, and to add the log events into those streams and then describe them. I've also added another statement that allows you to put metrics, which is required if you want to push custom monitor metrics.

6. Choose **Review Policy**.
7. On **Review Policy**, type `lambda-cloud-write` for the name.
8. Choose **Create Policy**.

Creating the Lambda IAM role

Now that we have two IAM policies, we will create a new Lambda IAM role and attach those two policies to it:

1. Sign in to the AWS Management Console and open the IAM console at `https://console.aws.amazon.com/iam/`
2. In the navigation pane, choose **Roles**
3. Choose **Create Role**
4. Select **AWS service** and, under that, select **Lambda**
5. Choose **Next: Permissions**
6. Under **Attach permissions policies | Filter polices**, type `dynamo-readonly-user-visits-api`
7. Select the checkbox for **dynamo-readonly-user-visits-api**
8. Under **Attach permissions policies | Filter polices**, type `lambda-cloud-write`
9. Select the checkbox for **lambda-cloud-write**
10. Choose **Next:Tags**
11. Choose **Next:Review**
12. On the **Review** page, type `lambda-dynamo-data-api` for the **Role name**
13. Choose **Create role**

You have created two IAM policies and attached them to a new Lambda execution role, which we will later associate with the Lambda function.

Creating and writing to a NoSQL database called DynamoDB using AWS

We are going to look at creating a DynamoDB table, writing data to the table from hardcoded values, writing data records from a file, and then we are going to show two different ways to query a table:

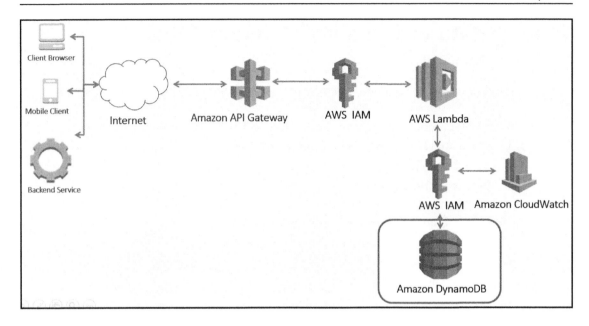

Creating a DynamoDB in AWS

The following steps show how to create a DynamoDB:

1. You need to sign in to the AWS Management Console first and then open the AWS DynamoDB console at `https://console.aws.amazon.com/dynamodb/`.
2. Choose **Create table** or, in the DynamoDB navigation pane, choose **Tables** and choose **Create table**.
3. In the **Create DynamoDB Table** window, perform the following steps:
    1. Under **Table name**, type `user-visits`
    2. In **Primary key** for **Partition key**, type `EventId` and choose **String**
    3. Check the **Add sort key** box
    4. In **Sort Key**, type `EventDay` and choose **Number**

The partition key and hash key can be used interchangeably, like sort key and range keys. A primary key can be the partition key alone, or a composite key with both a partition key and a sort key.

Writing data to DynamoDB using AWS

Perform the following steps:

1. Sign in to the AWS Management Console and open the DynamoDB console at `https://console.aws.amazon.com/dynamodb/`.
2. In the DynamoDB navigation pane, choose **Tables** and choose **user-visits**.
3. On the **user-visits** pane, choose the **Items*** tab.
4. Choose **Create Item**.
5. In the **Create Item** popup:
    1. Under **EventId String**, type `324`
    2. Under **EventDay Number**, type `20171001`
    3. Choose **+ > Append>Number**, for **field**, type `EventCount`, and for **Number**, type `3`
    4. Choose **Save**

You will now see a new record has been added in the **Items** tab in the lower-right pane, as the scan has also been done automatically.

DynamoDB is a managed NoSQL database, which means that each row can have different columns, and the names of the columns, known as **attributes**, are case-sensitive.

Querying DynamoDB using AWS

There are two types of searches you can do on DynamoDB; `Scan` and `Query`. `Scan` retrieves all the records from a table. `Query` makes use of the primary key to efficiently retrieve a set of records. Both allow you to have optional consistency, pagination, filters, conditions, and to choose the attributes to be returned. Generally speaking, `Scan` is useful if you want to retrieve all of the data, or the data across many different primary keys, and `Query` should be used when you have a primary key and want to retrieve all, or a filtered version of, the associated records.

DynamoDB Scan in AWS Management Console

Perform the following steps:

1. Sign in to the AWS Management Console and open the DynamoDB console at `https://console.aws.amazon.com/dynamodb/`

2. In the **DynamoDB** navigation pane, choose **Tables** and choose **user-visits**
3. On the **user-visits** pane choose the **Items*** tab
4. Select **Scan** from the dropdown
5. Optionally, choose **+Add Filter** to filter the query results
6. Select **Start Search**

You should now see the results in a table in the lower-right pane, with the columns **EventId**, **EventDay**, and **EventCount**.

DynamoDB Query in AWS Management Console

Perform the following steps:

1. Sign in to the AWS Management Console and open the DynamoDB console at
   `https://console.aws.amazon.com/dynamodb/`
2. In the **DynamoDB** navigation pane, choose **Tables** and choose **user-visits**
3. On the **user-visits** pane, choose the **Items*** tab
4. Select **Query** from the dropdown
5. Under **Partition Key**, type `324`
6. Under **Sort Key**, select > and type `20171001`
7. Select **Start Search**

You will see that no results are returned, as we are looking for records with **EventDay** greater than `20171001` and there are none in the current table.

Modify the following to find the record:

1. Under **Sort Key**, select >= and type `20171001`
2. Select **Start Search**

You will now see that the record we added is visible, as it meets the query search criterion.

Modify the following to find the record:

1. Under **Sort Key**, select **between** and type `20170930` and `20171002`
2. Select **Start Search**

Here, we use the **between** condition to retrieve the records too.

This query flexibility gives you the ability to retrieve data at a very low latency. However, you will notice that the partition key for the condition expression is always fixed to =, and has to be provided to all `Query` operations—this is something common in many NoSQL databases. If you do not have, or know, the primary key, then you need to use `Scan` instead.

Deleting DynamoDB using AWS

Let's delete the table, as we are going to re-create it using Python. Perform the following steps:

1. Log on to the console at `https://console.aws.amazon.com/dynamodb/`
2. Choose **Tables** from the left-hand **DynamoDB** menu
3. Choose **user-visits**
4. Choose **Delete table**
5. Choose **Delete**

Creating and writing to a NoSQL database called DynamoDB using Python

Now that we understand how to create a table, add data, and query DynamoDB using the AWS Console, we will look at how we can do this using only Python code.

We recommend you use a Python **Integrated development environment (IDE)** such as Eclipse PyDev (`http://www.pydev.org/download.html`) or PyCharm (`https://www.jetbrains.com/pycharm/`). You do not need to use an IDE, but I would recommend that you do. If you really want to, you can use VI, for example, on Linux to actually edit your code. But using an IDE allows you, for example, to run debugging or set up unit testing locally and step through it, which makes it easier and more productive for development.

First create the table using `Boto3 https://boto3.readthedocs.io/` in Python. Run the code in the following section in PyCharm or your favorite text editor.

Creating a DynamoDB table using Python

Here is generic Python code to create a table. Create a Python script called
`dynamo_table_creation.py` with the following code:

```python
import boto3

def create_dynamo_table(table_name_value, enable_streams=False,
                        read_capacity=1,
                        write_capacity=1,
                        region='eu-west-1'):
    table_name = table_name_value
    print('creating table: ' + table_name)
    try:
        client = boto3.client(service_name='dynamodb',
                              region_name=region)
        print(client.create_table(TableName=table_name,
                                  AttributeDefinitions=[{'AttributeName':
'EventId',
                                                         'AttributeType':
'S'},
                                                        {'AttributeName':
'EventDay',
                                                         'AttributeType':
'N'}],
                                  KeySchema=[{'AttributeName': 'EventId',
                                              'KeyType': 'HASH'},
                                             {'AttributeName': 'EventDay',
                                              'KeyType': 'RANGE'},
                                             ],
ProvisionedThroughput={'ReadCapacityUnits': read_capacity,
'WriteCapacityUnits': write_capacity}))
    except Exception as e:
        print(str(type(e)))
        print(e.__doc__)


def main():
    table_name = 'user-visits'
    create_dynamo_table(table_name, False, 1, 1)


if __name__ == '__main__':
    main()
```

Rather than creating a DynamoDB table in the AWS Console, here, we are creating it using the Python SDK Boto3. `main()` calls the method called `create_dynamo_table()`, which takes various parameters associated with the table we are going to create, `table_name` being the first. Ignore the `enable_streams` parameter, which we will use later. The other two are linked to the initial read and write capacities. These will have a direct impact on the costs, along with the table size and the data retrieved. That is why I have set them to `1` by default. The region parameter should be your AWS region.

We then create a `boto3.client()`, which is a low-level client representing DynamoDB. We then use this to create a table using `client.create_table()`, passing in the parameters passed in to our `create_dynamo_table()`, along with the partition key name, `EventId`, with its data type, `String`, indicated by `S`, and sort key name, `EventDay`, with its data type number indicated as `N`. All other attributes will be optional and arbitrary.

> You will notice a change in key terminology in DynamoDB between the Management Console and Boto3 descriptions, but they are synonyms: `Partition key (AWS Console) = Hash key (Boto3)` and `Sort key (AWS Console) = Range key (Boto3)`.

Both together, as a composite key, are called a primary key.

Writing to DynamoDB using Python

The following code writes three records to DynamoDB. Create another file called `dynamo_modify_items.py` with the following Python code:

```python
from boto3 import resource


class DynamoRepository:
    def __init__(self, target_dynamo_table, region='eu-west-1'):
        self.dynamodb = resource(service_name='dynamodb',
                        region_name=region)
        self.target_dynamo_table = target_dynamo_table
        self.table = self.dynamodb.Table(self.target_dynamo_table)

    def update_dynamo_event_counter(self, event_name,
                event_datetime, event_count=1):
        return self.table.update_item(
            Key={
                'EventId': event_name,
                'EventDay': event_datetime
            },
```

```
            ExpressionAttributeValues={":eventCount": event_count},
            UpdateExpression="ADD EventCount :eventCount")


def main():
    table_name = 'user-visits'
    dynamo_repo = DynamoRepository(table_name)
    print(dynamo_repo.update_dynamo_event_counter('324', 20171001))
    print(dynamo_repo.update_dynamo_event_counter('324', 20171001, 2))
    print(dynamo_repo.update_dynamo_event_counter('324', 20171002, 5))


if __name__ == '__main__':
    main()
```

Here, we use Boto3's `resource()`, which is a higher-level service resource with the repository pattern. We abstract all the DynamoDB-related code in the `DynamoRepository()` class that instantiates as `dynamo_repo` with `table_name`. `self.dynamodb.Table()` creates a table resource based on `table_name`. That will be used later on when calling `update_dynamo_event_counter()` to update DynamoDB records.

In `self.table.update_item()`, I first declare a variable called `eventCount` using `ExpressionAttributeValues`. I'm using this in the DynamoDB advanced *Update Expressions* (`https://docs.aws.amazon.com/amazondynamodb/latest/developerguide/Expressions.UpdateExpressions.html`), which is one of my favorite features in DynamoDB. Why? Because not all NoSQL databases can do something similar without having something like a semaphore lock and having the clients do a retry. It performs the following three actions in one atomic statement, while circumventing possible concurrency violations at the cost of eventual consistency:

1. Reads records matching the given `EventId=event_name` and `EventDay=event_datetime`
2. Creates a new item if it doesn't exist, setting `EventCount=1`
3. If it does already exist, then it increments `EventCount` by `event_count`

The first function calls `dynamo_repo.update_dynamo_event_counter('324', 20171001)`, sets `EventCount` to 1; the second function call, `dynamo_repo.update_dynamo_event_counter('324', 20171001, 2)`, increments `EventCount` by 2, so that it's now 3. The third function call adds a new record, as the `EventCount` or primary key, is different.

Querying DynamoDB using Python

Now that we have created a table and added data, we just need to write some code to query it. This will form part of the code that will be used in the Lambda function later on.

Create a Python script called `dynamo_query_table.py` with the following code:

```python
import decimal
import json

from boto3 import resource
from boto3.dynamodb.conditions import Key


class DecimalEncoder(json.JSONEncoder):
    """Helper class to convert a DynamoDB item to JSON
    """
    def default(self, o):
        if isinstance(o, decimal.Decimal):
            if o % 1 > 0:
                return float(o)
            else:
                return int(o)
        return super(DecimalEncoder, self).default(o)


class DynamoRepository:
    def __init__(self, target_dynamo_table, region='eu-west-1'):
        self.dynamodb = resource(service_name='dynamodb',
region_name=region)
        self.dynamo_table = target_dynamo_table
        self.table = self.dynamodb.Table(self.dynamo_table)

    def query_dynamo_record_by_parition(self, parition_key,
        parition_value):
        try:
            response = self.table.query(
                KeyConditionExpression=
                Key(parition_key).eq(parition_value))
            for record in response.get('Items'):
                print(json.dumps(record, cls=DecimalEncoder))
            return

        except Exception as e:
            print('Exception %s type' % str(type(e)))
            print('Exception message: %s ' % str(e))

    def query_dynamo_record_by_parition_sort_key(self,
```

```
                partition_key, partition_value, sort_key, sort_value):
            try:
                response = self.table.query(
                    KeyConditionExpression=Key(partition_key)
                    .eq(partition_value)
                    & Key(sort_key).gte(sort_value))
                for record in response.get('Items'):
                    print(json.dumps(record, cls=DecimalEncoder))
                return

            except Exception as e:
                print('Exception %s type' % str(type(e)))
                print('Exception message: %s ' % str(e))
def main():
    table_name = 'user-visits'
    partition_key = 'EventId'
    partition_value = '324'
    sort_key = 'EventDay'
    sort_value = 20171001

    dynamo_repo = DynamoRepository(table_name)
    print('Reading all data for partition_key:%s' % partition_value)
    dynamo_repo.query_dynamo_record_by_parition(partition_key,
        partition_value)

    print('Reading all data for partition_key:%s with date > %d'
            % (partition_value, sort_value))
    dynamo_repo.query_dynamo_record_by_parition_sort_key(partition_key,
        partition_value, sort_key, sort_value)
if __name__ == '__main__':
    main()
```

As I did earlier, I've created the `DynamoRepository` class, which abstracts all interactions with DynamoDB, including the connection and querying of tables. The following are the two methods used for querying the table using DynamoDB's `self.table.query()`:

- The `query_dynamo_record_by_parition()` method, which queries for records by `partition_key`, also known as a hash key, which in this case is `EventId`. Here, we are using just the equality condition in the `query()`, shown as `KeyConditionExpression=Key(partition_key).eq(parition_value))`.

- The `query_dynamo_record_by_parition_sort_key()` method, which queries for records by `partition_key` and `sort_key`, also known as a **range key**, which in this case is the `EventDate`. Here, we are using just the equality condition and the greater than or equal condition in the `query()` as `KeyConditionExpression=Key(partition_key).eq(partition_value) & Key(sort_key).gte(sort_value))`. This gives you the additional flexibility of quickly filtering by specific date ranges, for example, to retrieve the last 10 days of event data to display in a dashboard.

We then parse the returned records from the queries and print them to the console. This JSON will be what the Lambda will return to API Gateway as a response in the next section.

Creating a Lambda to query DynamoDB

Now that we have the `security` and `user-visits` table set up with some data, and know how to write code to query that DynamoDB table, we will write the Lambda Python code.

Creating the Lambda function

Now we have the IAM role with two IAM policies attached, create the Lambda function itself. Here, we are creating a function from scratch, as we want to walk through the full details to deepen your understanding of what is involved in creating a serverless stack. The following diagram shows data API architecture involving CloudWatch, DynamoDB, IAM, and Lambda:

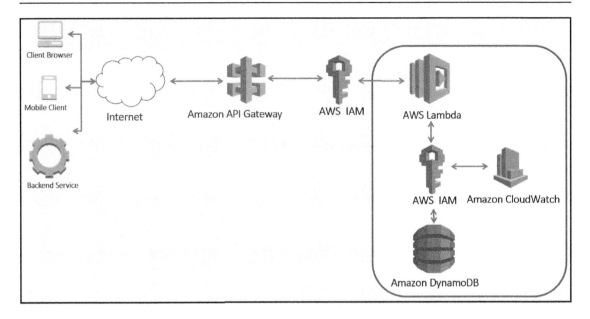

Perform the following steps:

1. Sign in to the AWS Management Console and open the AWS Lambda console at `https://console.aws.amazon.com/lambda/`.

2. Choose **Create function** or, in the AWS Lambda navigation pane, choose **Functions** and choose **Create function**.

3. On the **Create function** page, choose **Author from scratch** taking the following steps:
   1. For **Name**, type `lambda-dynamo-data-api`
   2. In **Runtime**, choose **Python 3.7**
   3. In **Role**, leave **Choose an existing role**
   4. In **Existing Role**, choose **lambda-dynamo-data-api**

4. Under the **Function Code**, a new file called `lambda_function.py` has been created. Copy and paste the following code under the **lambda_function** tab, overwriting the existing code:

```python
import json
import decimal

from boto3 import resource
from boto3.dynamodb.conditions import Key

class HttpUtils:
  def init(self):
    pass

@staticmethod
def parse_parameters(event):
    try:
        return_parameters =
        event['queryStringParameters'].copy()
    except Exception:
        return_parameters = {}
    try:
        resource_id = event.get('path', '').split('/')[-1]
        if resource_id.isdigit():
            return_parameters['resource_id'] = resource_id
        else:
            return {"parsedParams": None, "err":
                Exception("resource_id not a number")}
    except Exception as e:
        return {"parsedParams": None, "err": e}
        # Generally bad idea to expose exceptions
    return {"parsedParams": return_parameters, "err": None}

@staticmethod
def respond(err=None, err_code=400, res=None):
    return {
        'statusCode': str(err_code) if err else '200',
        'body': '{"message":%s}' % json.dumps(str(err))
            if err else
        json.dumps(res, cls=DecimalEncoder),
        'headers': {
            'Content-Type': 'application/json',
            'Access-Control-Allow-Origin': '*'
        },
    }

@staticmethod
def parse_body(event):
```

```
    try:
        return {"body": json.loads(event['body']),
                "err": None}
    except Exception as e:
        return {"body": None, "err": e}

class DecimalEncoder(json.JSONEncoder): def default(self, o):
if isinstance(o, decimal.Decimal): if o % 1 > 0:
    return float(o)
else: return int(o) return super(DecimalEncoder,
    self).default(o)

class DynamoRepository: def init(self, table_name):
self.dynamo_client = resource(service_name='dynamodb',
region_name='eu-west-1') self.table_name =
    table_name self.db_table =
self.dynamo_client.Table(table_name)

def query_by_partition_and_sort_key(self,
    partition_key, partition_value,
    sort_key, sort_value):
    response = self.db_table.query(KeyConditionExpression=
            Key(partition_key).eq(partition_value)
            & Key(sort_key).gte(sort_value))

    return response.get('Items')

def query_by_partition_key(self, partition_key,
    partition_value):
    response = self.db_table.query(KeyConditionExpression=
        Key(partition_key).eq(partition_value))
    return response.get('Items')

def print_exception(e): try: print('Exception %s type' %
str(type(e))) print('Exception message: %s '
                    % str(e)) except Exception: pass

class Controller(): def init(self): pass

@staticmethod
def get_dynamodb_records(event):
    try:
        validation_result = HttpUtils.parse_parameters(event)
        if validation_result.get(
        'parsedParams', None) is None:
            return HttpUtils.respond(
                err=validation_result['err'], err_code=404)
        resource_id = str(validation_result['parsedParams']
```

```
                            ["resource_id"])
            if validation_result['parsedParams']
                .get("startDate") is None:
                result = repo.query_by_partition_key(
                        partition_key="EventId",
                        partition_value=resource_id)
        else:
                start_date =
    int(validation_result['parsedParams']
                    ["startDate"])
                result = repo.query_by_partition_and_sort_key(
                        partition_key="EventId",
                        partition_value=resource_id,
                        sort_key="EventDay",
                        sort_value=start_date)
                        return HttpUtils.respond(res=result)

            except Exception as e:
            print_exception(e)
        return HttpUtils.respond(err=Exception('Not found'),
            err_code=404)

    table_name = 'user-visits' repo =
            DynamoRepository(table_name=table_name)

    def lambda_handler(event, context): response =
    Controller.get_dynamodb_records(event) return response
```

5. Choose **Save**.

A Lambda function always has a `handler` (https://docs.aws.amazon.com/lambda/latest/dg/python-programming-model-handler-types.html) and the main idea `event` is passed in, which contains the event source data. Here, this will be an API Gateway GET request. The other parameter is the `context` (https://docs.aws.amazon.com/lambda/latest/dg/python-context-object.html), which gives you details such as memory or the time-to-live for the Lambda.

You will recognize `class DynamoRepository()` from the earlier example, which deals with connection and queries. The new `HttpUtils` class is a set of utility methods for parsing the query string and body, and returning a response. The other new `Controller()` class controls the main flow. Here, it assumes the API Gateway request is a GET method, so it call functions to parse the request, query DynamoDB, and return the response to API Gateway.

Exception flows are built defensively, so all exceptions are caught (generally, best practice is to only catch specific named exceptions and raise the rest) rather than raised. This is because we want the API to be able to return a 4XX or 5XX in the event of exceptions. For debugging purposes, we are returning the exception too. In a production environment, you would not return the exception, only log it, as it could expose vulnerabilities in the code. As an analogy, you might remember seeing those SQL Server errors with a source error and full stack trace in yellow on a white background, from a poorly secured web application in your browser.

I recommend that you develop your Lambda code locally, but AWS has recently acquired an online IDE editor called Cloud9, which is where you have pasted the Python code, as shown in the following screenshot:

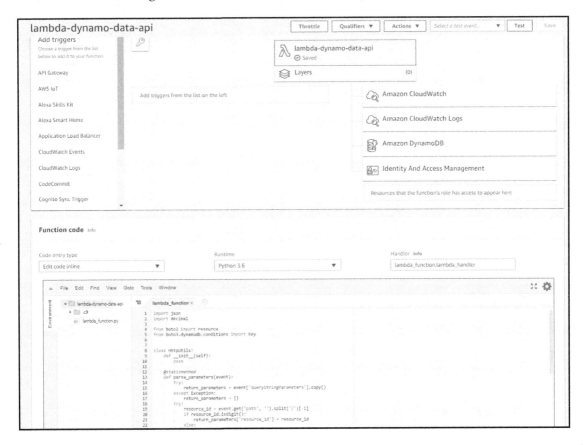

Testing the Lambda function

Now that we have the Lambda code deployed, we can test that it is working correctly in the AWS Management Console. Perform the following steps:

1. Sign in to the AWS Management Console and open the AWS Lambda console at `https://console.aws.amazon.com/lambda/`.

2. In the AWS Lambda navigation pane, choose **Functions**.

3. Choose **lambda-dynamo-data-api**.

4. Choose **Test**.

5. In **Configure test event**, under **Event name**, type `requestApiGatewayGetValid` and type or copy and paste the following code in the JSON document, overwriting the old code:

```
{
  "body": "{"test":"body"}",
  "resource": "/{proxy+}",
  "requestContext": {
    "resourceId": "123456",
    "apiId": "1234567890",
    "resourcePath": "/{proxy+}",
    "httpMethod": "GET",
    "requestId": "c6af9ac6-7b61-11e6-9a41-93e8deadbeef",
    "accountId": "123456789012",
    "identity": {
      "apiKey": null,
      "userArn": null,
      "cognitoAuthenticationType": null,
      "caller": null,
      "userAgent": "Custom User Agent String",
      "user": null, "cognitoIdentityPoolId": null,
      "cognitoIdentityId": null,
      "cognitoAuthenticationProvider": null,
      "sourceIp": "127.0.0.1",
      "accountId": null
    },
    "stage": "prod"
  },
  "queryStringParameters": {
    "StartDate": "20171009"
  },
  "headers": {
    "Via": "1.1 08f323deadbeefa7af34e5feb414ce27
            .cloudfront.net (CloudFront)",
    "Accept-Language": "en-US,en;q=0.8",
```

```
            "CloudFront-Is-Desktop-Viewer": "true",
            "CloudFront-Is-SmartTV-Viewer": "false",
            "CloudFront-Is-Mobile-Viewer": "false",
            "X-Forwarded-For": "127.0.0.1, 127.0.0.2",
            "CloudFront-Viewer-Country": "US", "Accept":
            "text/html,application/xhtml+xml,application/xml;q=0.9,
                image/webp,/;q=0.8",
            "Upgrade-Insecure-Requests": "1",
            "X-Forwarded-Port": "443", "Host": "1234567890
                .execute-api.us-east-1.amazonaws.com",
            "X-Forwarded-Proto": "https",
            "X-Amz-Cf-Id": "cDehVQoZnx43VYQb9j2-nvCh-
                9z396Uhbp027Y2JvkCPNLmGJHqlaA==",
            "CloudFront-Is-Tablet-Viewer": "false",
            "Cache-Control": "max-age=0",
            "User-Agent": "Custom User Agent String",
            "CloudFront-Forwarded-Proto": "https", "Accept-Encoding":
                "gzip, deflate, sdch"
        },
        "pathParameters": {
          "proxy": "path/to/resource"
        },
        "httpMethod": "GET",
        "stageVariables": {
          "baz": "qux"
        },
        "path": "/path/to/resource/324"
    }
```

6. The following are some important sections of the API Gateway GET request JSON:

   - The request uses the GET method from "httpMethod": "GET"
   - The resource or EventID is 324 and comes from "path": "/path/to/resource/324"
   - The query parameters come from "queryStringParameters": { "StartDate": "20171009"}

7. Choose **Create**.
8. Choose **Test** to run the test.

You should see from the execution results that the test succeeded with the sample API Gateway GET request. This includes the duration, memory used, and log output. Expand the details to see the response that will be sent to DynamoDB. It should be something like the following code:

```
{
   "statusCode": "200",
   "body": "[{"EventCount": 3, "EventDay": 20171001, "EventId":
          "324"}]",
   "headers": { "Content-Type": "application/json", "Access-Control-
          Allow-Origin": "*"
   }
}
```

If there is an error, check the **Log Output** for details, it could be something to do with the IAM role/policy, DynamoDB name, or the Lambda code.

Setting up the API Gateway and integrating it with a Lambda proxy

Now that we know the Lambda function works with some API Gateway test data, and returns a header and body with a statusCode of 200, we just need to add the API Gateway that will invoke the Lambda function, as shown in the following diagram:

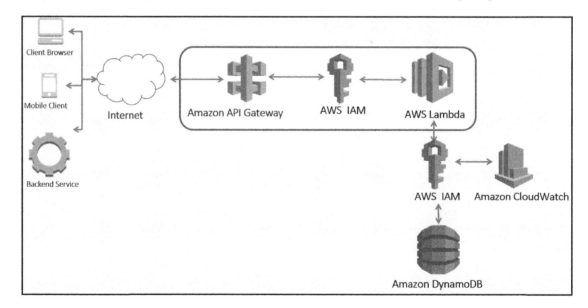

Perform the following steps:

1. Sign in to the AWS Management Console and open the API Gateway console at `https://console.aws.amazon.com/apigateway/`.
2. Choose **Get Started** or, in the Amazon API Gateway navigation pane, choose **APIs** and choose **Create API.**
3. On the **Create** page, perform the following steps:
    1. In **Choose Protocol**, select **REST**
    2. In **Create new API**, select **New API**
    3. Under Settings, type `metrics` for **API name**
    4. Choose `Regional` for **Endpoint Type**
    5. Choose **Create API**
4. Choose **Create Resource** from the **Actions** drop-down menu.
5. In the **New Child Resource** window, perform the following steps:
    1. In **Resource Name**, type `visits`
    2. In **Resource Path**, type `visits`
    3. Select **Enable API Gateway CORS**
    4. Choose **Create Resource.**
6. Select the `/visits` resource and choose **Create Resource** from the **Actions** drop-down menu.
7. In the **New Child Resource** window, perform the following steps:
    1. In **Resource Name**, type `{resourceId}`
    2. In **Resource Path**, type `{resourceId}`, replacing the default – resourceId– value
    3. Check **Enable API Gateway CORS**
    4. Choose **Create Resource**
8. Select the `/Vists/{resourceId}` resource and choose **Create Method** from the **Actions** drop-down menu.
9. Choose **GET** in the dropdown and then the checkmark to its right.
10. Choose the **GET** resource method in the **/visits/{resourceId} - GET - Setup** window:
    1. In **Integration type**, choose **Lambda Function**
    2. Check **Use Lambda Proxy integration**
    3. In **Lambda Region**, select your region from the dropdown

4. In **Lambda Function,** type `lambda-dynamo-data-api`
5. Check **Use Default Timeout**
6. Choose **Save**

11. Choose **OK** in the **Add Permission to Lambda Function.** This will allow API Gateway to invoke the Lambda function.

You should now have an API Gateway **GET - Method Execution** that looks like the following screenshot:

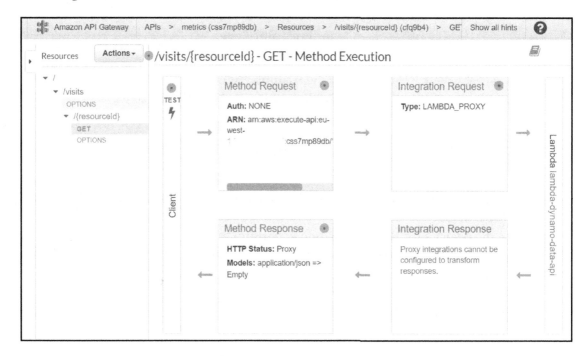

Finally do a quick test to make sure it works before we deploy it by performing the following steps:

1. Choose `/visits/{resourceId}` – `GET` in the **Resources** menu on the left
2. Choose **Test**
3. Type `324` under **Path {resourceId}**
4. Choose **Test**

You should see the status 200, latency, logs, and JSON response body, as shown in the following code:

```
[
    {
        "EventCount": 3,
        "EventDay": 20171001,
        "EventId": "324"
    },
    {
        "EventCount": 5,
        "EventDay": 20171002,
        "EventId": "324"
    }
]
```

If you don't get a 2XX status code, then look at the logs, which will help you diagnose the issue. It will probably be linked to security IAM roles.

Connecting API Gateway, Lambda, and DynamoDB

Now that we know the API Gateway integration with the Lambda function works, we will deploy it and get the URL. The architecture is shown in the following diagram:

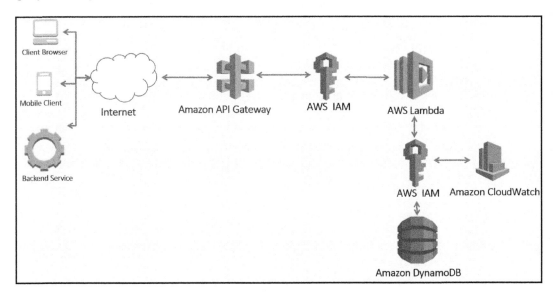

The workings of this architecture is as follows:

1. Sign in to the AWS Management Console and open the API Gateway console at `https://console.aws.amazon.com/apigateway/`.
2. In the Amazon API Gateway navigation pane, choose **APIs** and **metrics**.
3. Select **Resources** under **metrics** and `/Vists/{resourceId}`, and choose **Deploy API** from the **Actions** drop-down menu.
4. In the **Deploy API** pop-up window, perform the following steps:
   1. In **Deployment stage**, choose **[New Stage]**
   2. In **Stage name**, type `prod`
   3. In **Stage description**, type `prod`
   4. Select `Deploy`
5. The **Stages** under **metrics** should be automatically selected on the left-hand menu.
6. Select **GET** under `prod/visits/{resourceId}/GET` to get the invoke URL. The invoke URL should look like this: `https://{restapi_id}.execute-api.{region}.amazonaws.com/prod/visits/{resourceId}`.
7. Open a new browser tab and paste in the invoke URL: `https://{restapi_id}.execute-api.{region}.amazonaws.com/Prod/visits/{resourceId}`:
   - The response body will be `{"message": "Internal server error"}`
   - This is because we validated `resource_id` in the URL `parse_parameters()` function before querying DynamoDB to make sure that it is a number

8. Open a new browser tab and paste in the invoke URL: `https://{restapi_id}.execute-api.{region}.amazonaws.com/prod/visits/324`. As we have used the correct `resourceId` or `EventId` internally, you should see the following code in your browser tab:

```
[{
    "EventCount": 3,
    "EventDay": 20171001,
    "EventId": "324"
},
{
    "EventCount": 5,
    "EventDay": 20171002,
```

```
        "EventId": "324"
}]
```

9. Open a new browser tab and paste in the invoke
URL: `https://{restapi_id}.execute-api.{region}.amazonaws.com/Pr od/visits/324?startDate=20171002`. As we have added
the `startDate=20171002` parameter, you should see the following code in your
browser tab:

```
[{"EventCount": 5, "EventDay": 20171002, "EventId": "324"}]
```

That is using the other `query_by_partition_and_sort_key()` method in the Lambda
with the `startDate`.

We now have a fully working serverless data API, with the ability to run different types of
queries on DynamoDB.

Cleaning-up

You will need to delete the resources manually. I recommend you use the AWS Console to
do so. Perform the following steps:

1. Deleting API Gateway:
    1. Log on to the Console at `https://console.aws.amazon.com/ apigateway/`
    2. Choose **Resource** under **metrics** on the left-hand **APIs** menu
    3. Choose **Delete API** from the **Actions** drop-down menu
    4. Type metrics in the **Enter the name of the API before confirming this action** textbox
    5. Choose **Delete API**

2. Deleting the DynamoDB table:
    1. Log on to the Console at `https://console.aws.amazon.com/ dynamodb/`
    2. Choose **Tables** on the left-hand **DynamoDB** menu
    3. Choose **user-visits**
    4. Choose **Delete table**
    5. Choose **Delete**

3. Deleting the Lambda function:
    1. Log on to the Console at `https://console.aws.amazon.com/lambda/`
    2. Choose **Functions** on the left-hand **AWS Lambda** menu
    3. Choose **lambda-dynamo-data-api**
    4. Choose **Delete function** under the **Actions** menu
    5. Choose **Delete**

4. Deleting the IAM role and policies:
    1. Log on to the Console at `https://console.aws.amazon.com/iam/`
    2. Choose **Roles** in the IAM navigation pane
    3. Choose **lambda-dynamo-data-api**
    4. Choose **Delete Role** at the top-right
    5. Choose **Yes**
    6. Choose **Policies** in the IAM navigation pane
    7. Choose **Customer Managed** under **Filter Policies**
    8. Select the radio button next to **dynamo-readonly-user-visits**, and choose **Delete** under the **Policy actions** menu
    9. Choose **Delete** in the popup
    10. Select the radio button next to **lambda-cloud-write**, and choose **Delete** under the **Policy actions** menu
    11. Choose **Delete** in the popup

Summary

In this chapter, we have discussed security and why it is important. Applying the OWASP security by design principles is a good first step to ensure that your serverless stack is secure. We then discussed IAM roles and gave an overview of policies, explaining how they are the key documents to ensure restricted access to AWS resources. We then looked at an overview of some of the security concepts and principles regarding securing your serverless microservices, specifically regarding Lambda, API Gateway, and DynamoDB.

We then built a scalable serverless microservice with a RESTful data API. We started by creating a DynamoDB table, then added data to it, and queried it, first using the AWS console manually, then using the Python Boto3 SDK. We then built a simple Lambda to parse the request URL parameters, query DynamoDB, and return the records as part of a response body. We then looked at setting the integration between the Lambda and the API Gateway. We then connected everything together by deploying the API. We created a fully working API that is highly scalable, that you can tweak very easily for your own use cases, and that is very cost effective. In less than 30 minutes, you have created a highly scalable serverless microservice with an API. It is pay-per-usage for the API Gateway and the Lambda costs. For DynamoDB, you can actually change the read and write capacity very easily, set it to autoscale the read and write capacity based on the load, or even pay based on actual usage via the on-demand capacity mode, making it fully pay-per-API usage and data stored, avoiding traditional capacity planning or over-provisioning.

We have done a lot of work in the AWS Console, but in later chapters, we will be doing most of the work using the AWS CLI, or using code deployment pipelines. However, using the AWS Console should have given you a really good understanding of what you can do in AWS, and how Lambda integrates with DynamoDB and API Gateway. This solid foundation is really useful when we automate most of the creation and provisioning using configuration, scripts, and code. In the following chapters, we will be adding more functionality, automated testing, and deployment pipelines, and implementing microservice patterns.

In your organization, you will be developing a lot of source code and you won't want to deploy it manually, as we have done in this chapter. You will want to first test the code automatically to make sure it is working as expected, then you will deploy the stack in a repeatable fashion. This is needed for continuous integration or continuous delivery systems used in production.

In the next chapter, we are going to talk about how to deploy your serverless microservices using code and configuration to make the process more repeatable and scalable.

3
Deploying Your Serverless Stack

In the previous chapter, we created a fully-functional serverless data API using the API Gateway, Lambda, and DynamoDB with an IAM role, and tested it once it was deployed. However, most of the code and configuration was deployed manually; a process that is prone to error, not repeatable, and not scalable.

In this chapter, we are going to show you how to deploy all that infrastructure using only code and configuration. The topics covered are as follows:

- An overview of serverless-stack build and deploy options
- Creating a profile, an S3 bucket, IAM policies, and IAM roles resources
- Building and deploying with API Gateway, Lambda, and DynamoDB

An overview of serverless stack build and deploy options

In this section, we will discuss the challenges in manually provisioning infrastructure, infrastructure as code, building and deploying using the serverless application model, and building and deploying using the alternative options.

Manually provisioning infrastructure

The challenge of provisioning infrastructure is that it used to be a very manual process. For example, administrators would follow the steps described in a manual by clicking items on a user interface, running a set of commands, or logging into the servers and editing configuration files. This became more and more challenging with the growth of cloud computing and web frameworks that started to scale out. This could just be done with a monolithic architecture and their shared web servers or application servers. However, with the microservices architecture, there are different web servers and databases developed using different languages, and thousands of services running that need to be tested, built, and deployed independently.

There are a lot of efforts in deploying services manually in terms of cost, and also in terms of the ability to maintain such configurations at scale. The deployment of services has become slower to scale up and also to recover from any errors, as you would have the administrator, for example, remotely connecting via SSH onto your box, rebooting the machine, or trying to understand what the issues were and actually changing the configuration many times for many machines. It's also very difficult to test and make any process repeatable. Any configuration changes, done either by using a user interface or editing configuration files on one server, were not very repeatable and also prone to human error or misconfiguration. For example, we were using the AWS Management Console in the previous chapters. Had you made some errors in any of the configuration, you would have to go back, diagnose the issue, and fix it, which is very time-consuming.

In the next section, we will talk about infrastructure as code and how it helps to resolve the issues we have with manually provisioning infrastructure or deploying services.

Infrastructure as code

Infrastructure as code is basically the process of managing and provisioning your resources through definition files or code. It provides a centralized way to manage configuration in terms of implementation and version control. Suddenly, the resource management and provisioning becomes much more like the agile process in the systems-development life cycle in software. All changes are validated, tested, and provisioned as part of a release process and using standard deployment pipelines. This also provides the ability to copy configuration used to deploy infrastructure in one region to another.

For example, let's say you deployed infrastructure in the North Virginia region using code and configuration, you could easily modify it to make it work in the Ireland region too. This allows you to scale out very quickly in terms of your configuration around the infrastructure and this led to the development of the term DevOps. This is where developers get much more involved in the configuration, especially around the infrastructure, and the ops team (or operations teams) gets much more involved in the development process. The following diagram shows the different benefits of infrastructure as code:

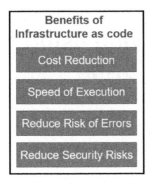

There are many benefits in using infrastructure as code. The first one is cost reduction, as you expend a lot less effort on simple and repetitive tasks. You can also reduce cost when scaling or deploying similar infrastructure.

When building any system, I usually like to build it so that it works with two environments. Once it works for two, it will work for many. For example, if you build code with a naming convention as a prefix or variable for each environment, such as **dev** for **development** and **stg** for **staging**, and substitute it when deploying it, then we can easily later add a **prd** prefix for **production**. Using a standard naming convention is always highly recommended. Another example could be to always have three characters as a convention or rule, so that you will not get into the situation of having multiple prefixes, such prod or production, that can introduce unexpected errors. In a config file, the environment variable that would get substituted could look like `${env}`.

The other point is the speed of execution; that is, your administrators or DevOps team can actually release infrastructure and services a lot faster than they would have done before. In addition, there's a reduction in the risk of errors that can be introduced, for example, through manual configuration or user-interface changes. Also, having traceability, validation, and testing at each step helps reduce the number of errors and issues. Overall, this helps reduce risks and improves security. Since you have this traceability, you can understand what was deployed and find out whether it was successful, or whether it's causing an issue and should be rolled back.

Building and deploying using the Serverless Application Model (SAM)

A tool that's emerged recently is the SAM (`https://github.com/awslabs/serverless-application-model`), which is maintained by AWS. It allows you to build and deploy your serverless stack. It provides a simplified way to define and deploy any serverless resources or applications. At its base, it employs a cloud formation, but using fewer lines of source code than if you used the AWS command line. The basic concept of using a SAM template file is that it can be either a JSON or YAML file that contains all the serverless configuration, and its mascot is SAM the squirrel.

Building and deploying using alternate options

There are alternative options to deploying your AWS serverless stack. The first one is the AWS **Command-Line Interface (CLI)**. The AWS CLI is an option when, for example, your organization does not want to use cloud formation stacks for everything or for parts of your serverless stack. The AWS CLI is also usually ahead of SAM in terms of feature releases. So, in this book, I use some commands to complement what is not built into SAM yet.

Serverless Framework, initially called JAWS, is built using Node.js technology. It was ahead of its time when it was first released, but now with the AWS SAM, it's an additional layer on top of AWS that's maintained by a third party. However, it does allow you to use other functions from other cloud providers, such as Google and Azure, which is a great feature, but I personally question the reuse of your function code across cloud providers as the event source, security, and data shape are different anyway.

Chalice and Zappa are Python-based frameworks for AWS and are similar to Python Flask and Bottle micro web frameworks, but again, they are another abstraction on top of AWS. You need to wait for any improvements to cascade through.

In addition, there's also the risk of having a dependency on those frameworks when AWS features are deprecated. You will need to keep in sync with them or rely on those other frameworks' open source committers to actually make changes or contribute directly. If I had to go for one, I would choose SAM, but I do accept that some people prefer serverless.

SAM needs an S3 bucket for package deployment, and the Lambda needs IAM policies and IAM roles. So let's look at that next.

Creating a profile, an S3 bucket, IAM policies, and IAM roles resources

We will first set up an S3 bucket that will hold the source code for the Lambda deployment package. IAM policies and roles allow API Gateway to invoke Lambda, and Lambda to access DynamoDB. We set them up using the AWS Management Console; here, we will use the AWS CLI and SAM.

The code, shell scripts, and configuration files used in this chapter are available under the `./serverless-microservice-data-api/` folder.

Creating an AWS credentials profile

Follow these steps to create an AWS credentials profile:

1. Create an AWS profile called demo:

```
$ aws configure --profile demo
```

2. Re-enter the same AWS `aws_access_key_id` and `aws_secret_access_key` details as in `Chapter 1`, *Serverless Microservices Architectures and Patterns*, for `newuser`.

   Alternatively, you can copy the `[default]` profile by copying `[default]` and creating a new entry for `[demo]`, as follows:

```
$ vi ~/.aws/credentials
[default]
aws_access_key_id =
AAAAAAAAAAAAAAAAAAAA
aws_secret_access_key =
11111111111111111111111111111111111111111

[demo]
aws_access_key_id =
AAAAAAAAAAAAAAAAAAAA
aws_secret_access_key =
11111111111111111111111111111111111111111
```

The code provided with this book needs a profile name (here, `demo`) to make use of the right keys; please change this in each shell script, `common-variables.sh`, for each project if you use another profile name.

Creating an S3 bucket

To deploy the Lambda source code, you will need to use an existing S3 bucket or create a new one—use the following code to create one:

```
$ aws s3api create-bucket --bucket <you-bucket-name> --profile demo --
create-bucket-configuration LocationConstraint=<your-aws-region> --region
<your-aws-region>
```

Ensure that `<your-bucket-name>` can be addressable—it must follow DNS naming conventions. To choose your AWS Region, refer to **AWS Regions and Endpoints** (`https://docs.aws.amazon.com/general/latest/gr/rande.html`). Generally, those in the USA can use `us-east-1` and those in Europe can use `eu-west-1`.

Setting up the configuration files for your AWS account

I've created a configuration file called `common-variables.sh` for each serverless project under `./bash/`, which creates environment variables that are used by the AWS CLI and SAM. You will need to modify them with your AWS account details. This is done to lay the groundwork to support multiple AWS accounts in more than one region. Here is an example of `common-variables.sh`:

```sh
#!/bin/sh
export profile="demo"
export region="<your-aws-region>"
export aws_account_id=$(aws sts get-caller-identity --query 'Account' --
profile $profile | tr -d '\"')
# export aws_account_id="<your-aws-accountid>"
export template="lambda-dynamo-data-api"
export bucket="<you-bucket-name>"
export prefix="tmp/sam"

# Lambda settings
export zip_file="lambda-dynamo-data-api.zip"
export files="lambda_return_dynamo_records.py"
```

Let's try to understand the code:

- Update `<your-aws-region>` with your AWS region, such as `us-east-1`.
- I'm dynamically determining the `aws_account_id`, but you can also hardcode it as shown in the comments, in which case uncomment the line and update `<your-aws-accountid>` with your AWS account ID. If you do not know it, you can find your account number in the **AWS Management Console | Support | Support Center** screen.
- `template` is the name of the SAM template that we will be using.
- `bucket` and `prefix` define the location of the deployed Lambda package.

Updating the polices and assuming roles files

You will need to change the AWS `aws_account_id` (currently set to `000000000000`) in the IAM policy documents stored under the `./IAM` folder. In addition, the region currently set to `eu-west-1` will have to be changed.

To replace your `aws_account_id` (assuming that your AWS `aws_account_id` is `111111111111`), you can do it manually or you can run the following command:

```
$ find ./ -type f -exec sed -i '' -e 's/000000000000/111111111111/' {} \;
```

Creating the IAM roles and policies

We created the IAM policies and roles manually in the AWS Management Console. We will now look at how we can create these using the AWS CLI.

Here is a JSON policy, `dynamo-readonly-user-visits.json`, that we have created under the `./IAM/` directory:

```
{
    "Version": "2012-10-17",
    "Statement": [
        {
            "Effect": "Allow",
            "Action": [
                "dynamodb:BatchGetItem",
                "dynamodb:DescribeTable",
                "dynamodb:GetItem",
                "dynamodb:Query",
                "dynamodb:Scan"
            ],
            "Resource": [
                "arn:aws:dynamodb:eu-west-1:000000000000:
                table/user-visits",
                "arn:aws:dynamodb:eu-west-1:000000000000:
                table/user-visits-sam"
            ]
        }
    ]
}
```

To summarize the policy, it says that we have `Query` and `Scan` access to two DynamoDB tables called `user-visits` that we created manually or in Python, and `user-visits-sam` that we are going to create in this chapter using SAM.

Create a policy that allows the Lambda function to write the logs to CloudWatch logs. Create a `lambda-cloud-write.json` file with the following content:

```json
{
  "Version": "2012-10-17",
  "Statement": [
    {
      "Effect": "Allow",
      "Action": [
        "logs:CreateLogGroup",
        "logs:CreateLogStream",
        "logs:PutLogEvents",
        "logs:DescribeLogStreams"
      ],
      "Resource": [
        "arn:aws:logs:*:*:*"
      ]
    },
    {
      "Effect": "Allow",
      "Action": [
        "cloudwatch:PutMetricData"
      ],
      "Resource": "*"
    }
  ]
}
```

When creating an IAM role, you also need specify the type of IAM role it can assume. We have created an `assume-role-lambda.json` file, which is known as a trusted entity:

```json
{
  "Version": "2012-10-17",
  "Statement": [
    {
      "Sid": "",
      "Effect": "Allow",
      "Principal": {
        "Service": "lambda.amazonaws.com"
      },
      "Action": "sts:AssumeRole"
    }
  ]
}
```

Having the preceding defined as JSON code allows us to version-control the security and permissions in AWS. In addition, if someone deleted them by mistake, we can simply recreate them in AWS.

We will now created a shell script called `create-role.sh`, under the `./bash` folder, to create a Lambda IAM role and three IAM policies, and attach them to the IAM role:

```sh
#!/bin/sh
#This Script creates a Lambda role and attaches the policies

#import environment variables
. ./common-variables.sh

#Setup Lambda Role
role_name=lambda-dynamo-data-api
aws iam create-role --role-name ${role_name} \
    --assume-role-policy-document file://../../IAM/assume-role-lambda.json \
    --profile $profile || true

sleep 1
#Add and attach DynamoDB Policy
dynamo_policy=dynamo-readonly-user-visits
aws iam create-policy --policy-name $dynamo_policy \
    --policy-document file://../../IAM/$dynamo_policy.json \
    --profile $profile || true

role_policy_arn="arn:aws:iam::$aws_account_id:policy/$dynamo_policy"
aws iam attach-role-policy \
    --role-name "${role_name}" \
    --policy-arn "${role_policy_arn}"  --profile ${profile} || true

#Add and attach cloudwatch_policy
cloudwatch_policy=lambda-cloud-write
aws iam create-policy --policy-name $cloudwatch_policy \
    --policy-document file://../../IAM/$cloudwatch_policy.json \
    --profile $profile || true

role_policy_arn="arn:aws:iam::$aws_account_id:policy/$cloudwatch_policy"
aws iam attach-role-policy \
    --role-name "${role_name}" \
    --policy-arn "${role_policy_arn}"  --profile ${profile} || true
```

Execute the script using `./create-role.sh`. It will create one IAM role and three IAM policies, and attach them to the IAM role. Notice that here code is idempotent on purpose, as policy changes need to be managed carefully as they could impact others.

Note that there is also the ability to create IAM roles in a SAM template, but using the AWS CLI means that the roles and policies can be reused rather than deleted when the serverless stack is deleted. This adds version control if you check them into the Git standard naming convention and helps the support team by centralizing the creation.

Checking the IAM roles and policies

The AWS CLI gives you feedback on the creation of the IAM role and policies, but you can also check in the AWS Management Console:

1. Sign in to the AWS Management Console and open the IAM console at `https://console.aws.amazon.com/iam/`.
2. In the IAM navigation pane, choose **Roles.**
3. Choose `lambda-dynamo-data-api` from the list of roles.
4. Choose **Show more** under **Permissions policies** on the **Permissions** tab.

You should see the following three attached policies:

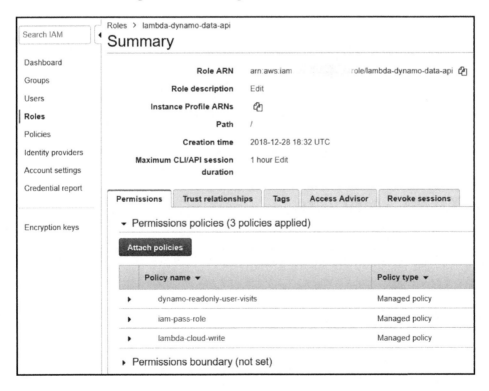

Building and deploying with API Gateway, Lambda, and DynamoDB

There are three steps involved in deploying a serverless stack:

1. Build Lambda as a ZIP package
2. Package your serverless stack using SAM and CloudFormation
3. Deploy your serverless stack using SAM and CloudFormation

Building the Lambda as a ZIP package

Install ZIP if it is not installed already. For Ubuntu/Debian, you can use `sudo apt-get install zip -y`. Create a file called `create-lambda-package.sh` with the following content:

```
#!/bin/sh
#setup environment variables
. ./common-variables.sh

#Create Lambda package and exclude the tests to reduce package size
(cd ../../lambda_dynamo_read;
mkdir -p ../package/
zip -FSr ../package/"${zip_file}" ${files} -x *tests/*)
```

This will create a ZIP file of the Lambda code only if the source code has changed. This is what will be deployed to AWS and there are advantages in separating these commands when we need to package third-party libraries.

SAM YAML template

We will use a SAM template to create the serverless stack. SAM uses YAML or JSON, and allows you to define the Lambda function and API Gateway settings, as well as create a DynamoDB table. The template looks as follows:

```
AWSTemplateFormatVersion: '2010-09-09'
Transform: 'AWS::Serverless-2016-10-31'
Description: >-
  This Lambda is invoked by API Gateway and queries DynamoDB.
Parameters:
    AccountId:
        Type: String
```

```
Resources:
  lambdadynamodataapi:
    Type: AWS::Serverless::Function
    Properties:
      Handler: lambda_return_dynamo_records.lambda_handler
      Runtime: python3.6
      CodeUri: ../../package/lambda-dynamo-data-api.zip
      FunctionName: lambda-dynamo-data-api-sam
      Description: >-
        This Lambda is invoked by API Gateway and queries DynamoDB.
      MemorySize: 128
      Timeout: 3
      Role: !Sub 'arn:aws:iam::${AccountId}:
                  role/lambda-dynamo-data-api'
      Environment:
        Variables:
          environment: dev
      Events:
        CatchAll:
          Type: Api
          Properties:
            Path: /visits/{resourceId}
            Method: GET
  DynamoDBTable:
    Type: AWS::DynamoDB::Table
    Properties:
      TableName: user-visits-sam
      SSESpecification:
        SSEEnabled: True
      AttributeDefinitions:
        - AttributeName: EventId
          AttributeType: S
        - AttributeName: EventDay
          AttributeType: N
      KeySchema:
        - AttributeName: EventId
          KeyType: HASH
        - AttributeName: EventDay
          KeyType: RANGE
      ProvisionedThroughput:
        ReadCapacityUnits: 1
        WriteCapacityUnits: 1
```

From top to bottom, we first specify the template type, a description, and pass in a string parameter, `AccountId`. We then specify the Lambda details, such as `Handler`, which is the entry point, location of the ZIP code, and give the function a name and description. We then choose 128 MB RAM as this is a proof of concept and we won't need more memory; we specify 3 for the timeout. After this, the Lambda will terminate even if it is still running; this limits costs and is reasonable, since we expect a synchronous response. We then have the IAM Lambda execution role with the `${AccountId}` parameter that gets passed in when we deploy the serverless stack.

We saw how to add the environment variable that will be available in the Lambda function. The variable is `environment: dev`.

We then have the trigger or event source for the Lambda function. Here, we create an API Gateway with a resource in the `/visits/{resourceId}` path with the `GET` method that will invoke a Lambda function with `resourceId`, which will be the `EventId`.

Finally, we create a DynamoDB table with an `EventId` hash of the data type `string` and an `EventDay` range of data type `number` using Python. To keep costs down (or free), I've put the read and write capacities to 1.

So in one SAM YAML file, we have configured the Lambda, API Gateway with its Lambda integration, and created a new DynamoDB table.

> For DynamoDB, I strongly recommend that you append `sam` at the end when it is a resource created by SAM, so you know the origin. I also recommend that if a DynamoDB table is shared between services, you create it using Boto3 or the AWS CLI. This is because the deletion of one serverless stack could mean the table is deleted for all services.

Packaging and deploying your serverless stack

Once the IAM role with policies, the ZIP package with the Lambda code, and SAM template are all created, you just need to run two CloudFormation commands to package and deploy your serverless stack.

The first command packages the Lambda code with the SAM template and pushes it to S3:

```
$ aws cloudformation package --template-file $template.yaml \
    --output-template-file ../../package/$template-output.yaml \
    --s3-bucket $bucket --s3-prefix backend \
    --region $region --profile $profile

Successfully packaged artifacts and wrote output template to file
```

```
../../package/lambda-dynamo-data-api-output.yaml.
Execute the following command to deploy the packaged template
aws cloudformation deploy --template-file /mnt/c/serverless-microservice-
data-api/package/lambda-dynamo-data-api-output.yaml --stack-name <YOUR
STACK NAME>
```

The second command deploys it to AWS:

```
$ aws cloudformation deploy --template-file ../../package/$template-
output.yaml \
    --stack-name $template --capabilities CAPABILITY_IAM \
    --parameter-overrides AccountId=${aws_account_id} \
    --region $region --profile $profile

Waiting for changeset to be created..
Waiting for stack create/update to complete
Successfully created/updated stack - lambda-dynamo-data-api
```

One of the great features in SAM is the ability to use parameters. Here, this is done when we deploy the stack with `--parameter-overrides AccountId=${aws_account_id}`. The benefit is that we can reuse the same SAM template for multiple environments, such as AWS Accounts and Regions, and any other parameters.

You can check that the stack has been deployed to AWS correctly by checking the AWS Management Console:

1. Sign into the AWS Management Console at `https://console.aws.amazon.com/cloudformation/`.
2. Choose **Management & Governance | CloudFormation** or search for **CloudFormation** under **Find services.**
3. In the CloudFormation pane, choose **lambda-dynamo-data-api**.
4. Choose **Events**. This shows the different events and is very useful for diagnosing any issues you get when deploying a stack. Usually, it will be a naming conflict (for example, a **DynamoDB** table with the same name exists) or an **IAM**-related issue (for example, a role does not exist).

5. Choose **Resources**. This shows the resources that are managed by this **CloudFormation** stack:

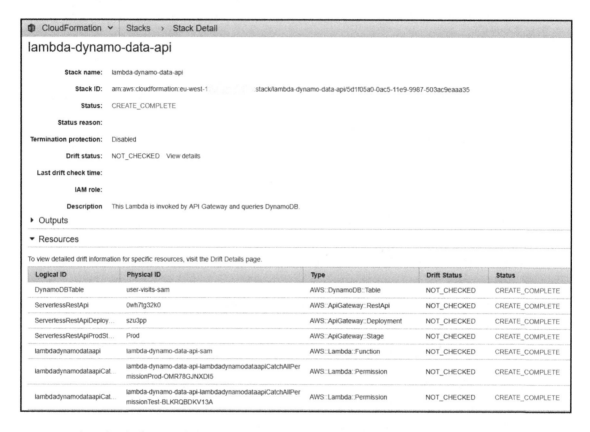

You can also check the AWS Management Console directly if the API Gateway, Lambda function, and DynamoDB table have been created correctly.

For example, here is the same Lambda we created using Python, but deployed and managed by SAM. Unless you are doing a proof of concept, it is recommended that any further changes are managed by configuration changes rather than changes in the AWS Management Console, as this would break the infrastructure as code and automation:

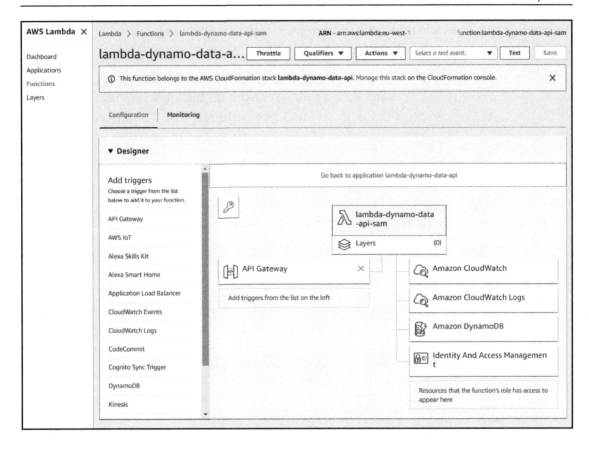

Putting it all together

In this chapter, we deployed a fully-working serverless stack without the need to use the AWS Management Console to configure any settings using the user interface. This is the recommended way to deploy infrastructure and code as it's much more repeatable, scalable, and less prone to errors. It also allows you to do things such as revert configurations when everything is version-controlled in Git.

The shell scripts available under the `./serverless-microservice-data-api/bash` folder:

- `common-variables.sh`: Environment variables used by other scripts
- `create-role.sh`: Lambda IAM role created with the three policies attached
- `lambda-dynamo-data-api.yaml`: Defines the SAM YAML template
- `create-lambda-package.sh`: Creates the Lambda ZIP package
- `build-package-deploy-lambda-dynamo-data-api.sh`: Orchestrates the building of the Lambda ZIP, packaging, and deployment

Here are the contents of `build-package-deploy-lambda-dynamo-data-api.sh`, which you can run when you modify your Lambda code or other SAM configuration settings:

```bash
#!/usr/bin/env bash

# Variables
. ./common-variables.sh

#Create Zip file of your Lambda code (works on Windows and Linux)
./create-lambda-package.sh

#Package your Serverless Stack using SAM + Cloudformation
aws cloudformation package --template-file $template.yaml \
    --output-template-file ../../package/$template-output.yaml \
    --s3-bucket $bucket --s3-prefix backend \
    --region $region --profile $profile

#Deploy your Serverless Stack using SAM + Cloudformation
aws cloudformation deploy --template-file ../../package/$template-output.yaml \
    --stack-name $template --capabilities CAPABILITY_IAM \
    --parameter-overrides AccountId=${aws_account_id} \
    --region $region --profile $profile
```

Manually testing the serverless microservice

The steps for testing are as follows:

1. Sign in to the AWS Management Console and open the API Gateway console at `https://console.aws.amazon.com/apigateway/`.
2. In the Amazon API Gateway navigation pane, choose **APIs | lambda-dynamo-data-api | Stages**.

3. Select **GET** under `Prod/visits/{resourceId}/GET` to get the invoke URL, which should look like `https://{restapi_id}.execute-api.{region}.amazonaws.com/Prod/visits/{resourceId}`.

4. Open a new browser tab and enter the `https://{restapi_id}.execute-api.{region}.amazonaws.com/Prod/visits/{resourceId}` URL. You will get the `{"message":"resource_id not a number"}` response body. This is because we validated `resource_id` in the `parse_parameters()` URL function before querying DynamoDB to make sure that it is a number.

5. Open a new browser tab and enter the `https://{restapi_id}.execute-api.{region}.amazonaws.com/Prod/visits/324` URL. As we have used the correct `resourceId`, you should see [ ] in your browser tab.

Why are we getting no data?

Well, no data has been loaded into the `user-visits-sam` DynamoDB table, that is why!

Run `python3 ./aws_dynamo/dynamo_modify_items.py` to load some records into the `user-visits-sam` DynamoDB table.

Here are the contents of `dynamo_modify_items.py`:

```python
from boto3 import resource

class DynamoRepository:
    def __init__(self, target_dynamo_table, region='eu-west-1'):
        self.dynamodb = resource(service_name='dynamodb',
region_name=region)
        self.target_dynamo_table = target_dynamo_table
        self.table = self.dynamodb.Table(self.target_dynamo_table)

    def update_dynamo_event_counter(self, event_name,
            event_datetime, event_count=1):
        return self.table.update_item(
            Key={
                'EventId': event_name,
                'EventDay': event_datetime
            },
            ExpressionAttributeValues={":eventCount": event_count},
            UpdateExpression="ADD EventCount :eventCount")

def main():
    table_name = 'user-visits-sam'
```

```
    dynamo_repo = DynamoRepository(table_name)
    print(dynamo_repo.update_dynamo_event_counter('324', 20171001))
    print(dynamo_repo.update_dynamo_event_counter('324', 20171001, 2))
    print(dynamo_repo.update_dynamo_event_counter('324', 20171002, 5))

if __name__ == '__main__':
    main()
```

Now go to the same endpoint in the browser, and you should get the following data back:

```
[{"EventCount": 3, "EventDay": 20171001, "EventId": "324"}, {"EventCount":
5, "EventDay": 20171002, "EventId": "324"}]
```

Open a new browser tab and enter
the `https://{restapi_id}.execute-api.{region}.amazonaws.com/Prod/visits/`
`324?startDate=20171002` URL. As we have added
the `startDate=20171002` parameter, you should see the following in your browser tab:

```
[{"EventCount": 5, "EventDay": 20171002, "EventId": "324"}]
```

Making code and configuration changes

Code rarely stays static, as new requirements and business requests arise. To illustrate how well changes are supported, assume we made some changes to the Lambda function Python code and now want to use Python 3.7, instead of Python 3.6.

We can update the code, configuration, and stack in three steps:

1. Change the `lambda_return_dynamo_records.py` Python code to make it compliant with Python 3.7.
2. Change the `lambda-dynamo-data-api.yaml` SAM template as follows:

   ```
   Resources:
     lambdadynamodataapi:
       Type: AWS::Serverless::Function
       Properties:
         Handler: lambda_return_dynamo_records.lambda_handler
         Runtime: python3.7
   ```

3. Run `./build-package-deploy-lambda-dynamo-data-api.sh`. This will rebuild the Lambda ZIP package, as the code has changed. Package and deploy the code and SAM configuration, and then CloudFormation will manage and deploy the changes.

Deleting the serverless stack

When you no longer need your serverless stack, you can delete it in the AWS Management Console under CloudFormation:

1. Sign in to the AWS Management Console and open the CloudFormation console at `https://console.aws.amazon.com/cloudformation/`.
2. From the list, select **lambda-dynamo-data-api**.
3. Choose **Actions** and then **Delete Stack.**
4. Choose **Yes, Delete** when prompted.

Alternatively, you can run the following shell script with `./delete-stack.sh`:

```
#!/usr/bin/env bash
. ./common-variables.sh
aws cloudformation delete-stack --stack-name $template --region $region --
profile $profile
```

Summary

You now have a much deeper understanding and some practical experience of manually deploying your serverless stack in a repeatable and consistent way using infrastructure-as-code principles. You can adapt these for your organization's serverless microservice needs. You know the service deployment options and you used the AWS CLI to create a bucket, IAM roles, and IAM policies, as well as the AWS SAM to deploy the API Gateway, Lambda, and DynamoDB. You also saw how you can easily modify the SAM template file to propagate changes throughout the stack. The full Python source code, IAM policies, roles, Linux, and shell scripts are provided with this book so you can adapt it for your needs. You can now take advantage of them without having to use the AWS management console GUI manually, and only need to modify the scripts when deploying other serverless microservices.

Now that we have shown you how to deploy the stacks, it's really important that you know the code is functioning and performing as expected, especially as the code base grows and will be used in a production environment. We have not yet covered the mechanism for automated deployment and testing. So, in the next chapter, we are going to discuss and walk through the different types of testing you should use on your serverless microservice.

4
Testing Your Serverless Microservice

In the previous chapter, we created a fully functional serverless data API using API Gateway, Lambda, and DynamoDB, and we deployed it to the AWS CLI. The testing we showed was performed in the AWS Management Console and browser, which is fine for small amounts of simple code development as a proof of concept, but is not recommended for development or production systems.

It is much more efficient for a developer to first develop and test locally, and it is essential for continuous delivery to have automated tests. This chapter is all about testing.

Testing could easily cover a whole book, but we will keep things very practical and focused on testing your serverless code and the data API we deployed in `Chapter 3`, *Deploying Your Serverless Stack*. This will include unit testing, mocking, local debugging, integration testing, running the Lambda or serverless APIs with an HTTP server locally in a Docker container, and load testing.

In this chapter, we will cover the following topics:

- Unit testing your Python Lambda code
- Running and debugging your AWS Lambda code locally
- Integration testing using real test data
- The AWS **Serverless Application Model (SAM)** CLI
- Loading and end-to-end testing at scale
- Strategies to reduce the API's latency
- Cleaning up

Unit testing your Python Lambda code

In this section, we are going to look at why testing is important, as well as the sample data we can use for testing, unit testing, and mocking.

Why is testing important?

Think about the **collaboration and teamwork** taking place in large, distributed teams of developers in different countries, and imagine they want to collaborate on the same source code repository and check code changes at different times. It's very important for these teams to understand the code and be able to test it locally to see how it works, whether their changes will impact existing services, and whether the code is still working as expected.

Testing is important to ensure we have **quality** in the delivery or user experience. By having lots of tests, you can identify defects early and fix them. For example, if there are major bugs detected, you could decide not to release the recent update and fix the issues before doing the release.

Another major point is **usability**. For example, your client might have performance or non-functional requirements. Imagine an e-commerce website, for example, where you could add items and would have to wait a whole minute for them to be added to the basket. In most cases, that would be unacceptable and the user would lose trust in the platform. Ideally, you would have a testing process to make sure that the latency is still low and the website is responsive. Other examples that require testing would be features that are not working as expected or user interface defects that prevent the users from completing the tasks they want to do.

It is important to have **shorter release cycles**. Using automation, you can run a thousand tests automatically and consistently, without needing a human to test different parts of the site manually, test the APIs manually, or rigorously inspect the code before any release. Before every release into production, you run those thousand tests, making you much more confident that everything is working as expected, and if you do spot an issue those thousand tests missed in production, you can fix it and add a new test for that scenario too.

Types of testing

Testing can be done manually, as we have done with the AWS Management Console, which is prone to error and not scalable. Generally, tests are automated using test suites, which are written in advance and are essential for continuous integration and continuous delivery.

There are many definitions and types of software testing available; it could take a whole book to cover them all. Here, we will focus on the three main ones that are relevant for our serverless stacks:

- **Unit testing:** The low-level testing of individual software modules in isolation, typically done by the developers and used in **test-driven development** (TDD). These types of tests are usually quick to execute.
- **Integration testing**: Verifies that all combined services after integration are working correctly. These are generally more expensive to run as many services need to be running.
- **Load testing**: This is non-functional testing used to check how the system performs under a heavy load. It is sometimes also called performance or stress testing as it helps to understand the availability and reliability of a platform.

Unit testing Lambda Python code

It is not easy to debug in AWS Management Console; it's much more constructive to debug code locally and later automate the process.

We know from previous chapters that the Lambda event source is an API Gateway GET request. As we are only looking at a subset of the data, this full JSON payload can be simulated with a few lines of Python code too.

The sample test data

Here, we have a test case with a setUp() method that is run once at the start of the test suite and a tearDown() method that is run at the end.

Here is a subset of the contents linked to the test setup and teardown at the top of `serverless-microservice-data-api/test/test_dynamo_get.py`:

```
import unittest
import json

class TestIndexGetMethod(unittest.TestCase):
    def setUp(self):
        self.validJsonDataNoStartDate = json.loads('{"httpMethod":
        "GET","path": "/path/to/resource/324","headers": ' \ 'null} ')
        self.validJsonDataStartDate =
        json.loads('{"queryStringParameters": {"startDate":
        "20171013"},' \ '"httpMethod": "GET","path": "/path/to/resource
        /324","headers": ' \ 'null} ')
        self.invalidJsonUserIdData =
        json.loads('{"queryStringParameters": {"startDate":
        "20171013"},' \ '"httpMethod": "GET","path": "/path/to/resource
        /324f","headers": ' \ 'null} ')
        self.invalidJsonData = "{ invalid JSON request!} "
    def tearDown(self):
        pass
```

I created four different JSON Python dictionaries:

- `self.validJsonDataNoStartDate`: A valid GET request with no StartDate filter
- `self.validJsonDataStartDate`: A valid GET request with a StartDate filter
- `self.invalidJsonUserIdData`: An invalid UserId that is not a number
- `self.invalidJsonData`: Invalid JSON that cannot be parsed

The unit test

Here are the units tests that can be found in the middle of `serverless-microservice-data-api/test/test_dynamo_get.py`:

```
    def test_validparameters_parseparameters_pass(self):
        parameters = lambda_query_dynamo.HttpUtils.parse_parameters(
                self.validJsonDataStartDate)
        assert parameters['parsedParams']['startDate'] == u'20171013'
        assert parameters['parsedParams']['resource_id'] == u'324'

    def test_emptybody_parsebody_nonebody(self):
        body = lambda_query_dynamo.HttpUtils.parse_body(
            self.validJsonDataStartDate)
        assert body['body'] is None
```

```
def test_invalidjson_getrecord_notfound404(self):
    result = lambda_query_dynamo.Controller.get_dynamodb_records(
            self.invalidJsonData)
    assert result['statusCode'] == '404'

def test_invaliduserid_getrecord_invalididerror(self):
    result = lambda_query_dynamo.Controller.get_dynamodb_records(
            self.invalidJsonUserIdData)
    assert result['statusCode'] == '404'
    assert json.loads(result['body'])['message'] ==
        "resource_id not a number"
```

I'm using a prefix of `test` so Python test suites can automatically detect them as unit tests, and I'm using the triplet unit test naming convention for the test methods: the method name, the state under test, and the expected behavior. The test methods are as follows:

- `test_validparameters_parseparameters_pass()`: Checks that the parameters are parsed correctly.
- `test_emptybody_parsebody_nonebody()`: We are not using a body in the GET method, so we want to make sure that it still works if none is provided.
- `test_invalidjson_getrecord_notfound404()`: Check how the Lambda will react with an invalid JSON payload.
- `test_invaliduserid_getrecord_invalididerror()`: Check how the Lambda will react to an invalid non-number `userId`.

The preceding does not query DynamoDB for the records. If we want to do so, we should have DynamoDB running, use the new DynamoDB Local (https://docs.aws.amazon.com/amazondynamodb/latest/developerguide/DynamoDBLocal.html), or we can mock the DynamoDB calls, which is what we will look at next.

Mocking

There is a Python AWS mocking framework called Moto (http://docs.getmoto.org/en/latest/), but I prefer to use a generic one called `mock`, which is much more widely supported in the Python community and from Python 3.3 is included in the Python standard library.

The following mocking code can be found at the bottom of `serverless-microservice-data-api/test/test_dynamo_get.py`:

```
from unittest import mock

    mock.patch.object(lambda_query_dynamo.DynamoRepository,
                      "query_by_partition_key",
                      return_value=['item'])
    def test_validid_checkstatus_status200(self,
        mock_query_by_partition_key):
        result = lambda_query_dynamo.Controller.get_dynamodb_records(
                self.validJsonDataNoStartDate)
        assert result['statusCode'] == '200'

    @mock.patch.object(lambda_query_dynamo.DynamoRepository,
                       "query_by_partition_key",
                       return_value=['item'])
    def test_validid_getrecord_validparamcall(self,
        mock_query_by_partition_key):
lambda_query_dynamo.Controller.get_dynamodb_records(
self.validJsonDataNoStartDate)
mock_query_by_partition_key.assert_called_with(
    partition_key='EventId',
    partition_value=u'324')

    @mock.patch.object(lambda_query_dynamo.DynamoRepository,
                       "query_by_partition_and_sort_key",
                       return_value=['item'])
    def test_validid_getrecorddate_validparamcall(self,
        mock_query_by_partition_and_sort_key):
            lambda_query_dynamo.Controller.get_dynamodb_records(
                self.validJsonDataStartDate)
mock_query_by_partition_and_sort_key.assert_called_with(partition_key='
    EventId',
    partition_value=u'324',
    sort_key='EventDay',
    sort_value=20171013)
```

The key observations from this code are as follows:

- `@mock.patch.object()` is a decorator for the `query_by_partition_key()` or `query_by_partition_and_sort_key()` method we are mocking from the `DynamoRepository()` class.
- `test_validid_checkstatus_status200()`: We mock the calls to `query_by_partition_key()`. If the query is valid, we get a `'200'` status code back.

- `test_validid_getrecords_validparamcall()`: We mock the calls to `query_by_partition_key()` and check the method is called with the correct parameters. Note that don't need to check that the lower-level `boto3` `self.db_table.query()` method works.
- `test_validid_getrecordsdate_validparamcall()`: We mock the calls to `query_by_partition_and_sort_key()` and check that the method is called with the correct parameters.

> You are not here to test existing third-party libraries or Boto3, but your code and integration with them. Mocking allows you to replace parts of the code under test with mock objects and make an assertion about the method or attributes.

Running the unit test

Now that we have all the test suites, rather than run them in your IDE, such as PyCharm, you can run the tests from the root folder using the following bash command:

```
$ python3 —m unittest discover test
```

`unittest` automatically detects that all of the test files must be modules or packages importable from the top-level directory of the project. Here, we just want to run the tests from the test folder that begin with the `test_` prefix.

I have created a shell script under `serverless-microservice-data-api/bash/apigateway-lambda-dynamodb/unit-test-lambda.sh`:

```
#!/bin/sh (cd ../..; python3 —m unittest discover test)
```

Code coverage

We won't discuss it in depth, but code coverage is another important measure used in software engineering. Code coverage measures the degree of code that your test suites cover. The main idea is that the higher the coverage percentage, the more code is covered by tests, so the less likely you are to create undetected bugs and the service should behave as intended. These reports can help developers come up with additional tests or scenarios to increase the coverage percentage.

Test-coverage-related Python packages include `coverage`, `nose`, and the more recent `nose2`, which can provide coverage reports. For example, you can run the following to get a test-coverage analysis report of your Lambda code with `nose` or `nose2`:

```
$ nosetests test/test_dynamo_get.py --with-coverage --cover-package
lambda_dynamo_read -v
$ nose2 --with-coverage
```

When we commence with writing our own tests, we have an option to use an additional set of tools to do so. Such tools are known as code coverage tools. Codecov and Coveralls are examples of such tools. When we want to analyze the code that is written via hosting services such as GitHub, these tools are very helpful as they provided a complete breakdown of whether the lines are tested.

Running and debugging your AWS Lambda code locally

Sometimes you want to simulate an API Gateway payload with a local Lambda against a real instance of remote DynamoDB hosted in AWS. This allows you to debug and build up unit tests with real data. In addition, we will see how these can later be used in the integration test.

Batch-loading data into DynamoDB

We will first discuss how to batch-load data into DynamoDB from a **comma-separated values (CSV)** file called `sample_data/dynamodb-sample-data.txt`. Rather than insert an individual statement for each item, this is a much more efficient process, as the data file is decoupled from the Python code:

```
EventId,EventDay,EventCount
324,20171010,2
324,20171012,10
324,20171013,10
324,20171014,6
324,20171016,6
324,20171017,2
300,20171011,1
300,20171013,3
300,20171014,30
```

Add another method, called `update_dynamo_event_counter()`, that updates DynamoDB records using the `DynamoRepository` class.

Here are the contents of the `serverless-microservice-data-api/aws_dynamo/dynamo_insert_items_from_file.py` Python script:

```python
from boto3 import resource

class DynamoRepository:
    def __init__(self, target_dynamo_table, region='eu-west-1'):
        self.dynamodb = resource(service_name='dynamodb',
region_name=region)
        self.target_dynamo_table = target_dynamo_table
        self.table = self.dynamodb.Table(self.target_dynamo_table)

    def update_dynamo_event_counter(self, event_name,
        event_datetime, event_count=1):
        response = self.table.update_item(
            Key={
                'EventId': str(event_name),
                'EventDay': int(event_datetime)
            },
            ExpressionAttributeValues={":eventCount":
                int(event_count)},
            UpdateExpression="ADD EventCount :eventCount")
        return response
```

Here, we have a `DynamoRepository` class that instantiates a connection to DynamoDB in `__init__()` and an `update_dynamo_event_counter()` method that updates the DynamoDB records if they exist, or adds a new one if they don't using the passed-in parameters. This is done in one atomic action.

Here's the second half of the `serverless-microservice-data-api/aws_dynamo/dynamo_insert_items_from_file.py` Python script:

```python
 import csv
table_name = 'user-visits-sam'
input_data_path = '../sample_data/dynamodb-sample-data.txt'
dynamo_repo = DynamoRepository(table_name)
with open(input_data_path, 'r') as sample_file:
    csv_reader = csv.DictReader(sample_file)
    for row in csv_reader:
        response = dynamo_repo.update_dynamo_event_counter(row['EventId'],
row['EventDay'],
row['EventCount'])
        print(response)
```

This Python code opens the CSV, extracts the header row, and parses each row while writing it to the DynamoDB table called `user-visits-sam`.

Now that we have loaded some data rows into the DynamoDB table, we will query the table by debugging a local Lambda function.

Running the Lambda locally

Here is a full example API Gateway request, `serverless-microservice-data-api/sample_data/request-api-gateway-valid-date.json`, that a proxy Lambda function would receive as an event. These can be generated by printing the real API Gateway JSON event, which the Lambda gets as an event source into CloudWatch logs:

```
{
  "body": "{\"test\":\"body\"}",
  "resource": "/{proxy+}",
  "requestContext": {
    "resourceId": "123456",
    "apiId": "1234567890",
    "resourcePath": "/{proxy+}",
    "httpMethod": "GET",
    "requestId": "c6af9ac6-7b61-11e6-9a41-93e8deadbeef",
    "accountId": "123456789012",
    "identity": {
      "apiKey": null,
      "userArn": null,
      "cognitoAuthenticationType": null,
      "caller": null,
      "userAgent": "Custom User Agent String",
      "user": null,
      "cognitoIdentityPoolId": null,
      "cognitoIdentityId": null,
      "cognitoAuthenticationProvider": null,
      "sourceIp": "127.0.0.1",
      "accountId": null
    },
    "stage": "prod"
  },
  "queryStringParameters": {
    "foo": "bar"
  },
  "headers": {
    "Via": "1.1 08f323deadbeefa7af34d5feb414ce27.cloudfront.net
            (CloudFront)",
    "Accept-Language": "en-US,en;q=0.8",
```

```
        "CloudFront-Is-Desktop-Viewer": "true",
        "CloudFront-Is-SmartTV-Viewer": "false",
        "CloudFront-Is-Mobile-Viewer": "false",
        "X-Forwarded-For": "127.0.0.1, 127.0.0.2",
        "CloudFront-Viewer-Country": "US",
        "Accept": "text/html,application/xhtml+xml,application/xml;
                  q=0.9,image/webp,*/*;q=0.8",
        "Upgrade-Insecure-Requests": "1",
        "X-Forwarded-Port": "443",
        "Host": "1234567890.execute-api.us-east-1.amazonaws.com",
        "X-Forwarded-Proto": "https",
        "X-Amz-Cf-Id": "cDehVQoZnx43VYQb9j2-nvCh-
                       9z396Uhbp027Y2JvkCPNLmGJHqlaA==",
        "CloudFront-Is-Tablet-Viewer": "false",
        "Cache-Control": "max-age=0",
        "User-Agent": "Custom User Agent String",
        "CloudFront-Forwarded-Proto": "https",
        "Accept-Encoding": "gzip, deflate, sdch"
    },
    "pathParameters":{
      "proxy": "path/to/resource"
    },
    "httpMethod": "GET",
    "stageVariables": {
      "baz": "qux"
    },
    "path": "/path/to/resource/324"
}
```

Rather than relying on another third-party framework for local debugging (such as the SAM CLI), you can debug a Lambda function directly by calling it with the JSON `Dict` event. This means that you don't need any additional libraries to run and it's native Python.

The contents of `serverless-microservice-data-api/test/run_local_api_gateway_lambda_dynamo.py` are an example of debugging a Lambda function locally with services such as DynamoDB in AWS:

```python
import json

from lambda_dynamo_read import lambda_return_dynamo_records as
lambda_query_dynamo

with open('../sample_data/request-api-gateway-valid-date.json', 'r') as
sample_file:
    event = json.loads(sample_file.read())
print("lambda_query_dynamo\nUsing data: %s" % event)
```

```
print(sample_file.name.split('/')[-1]) response =
lambda_query_dynamo.lambda_handler(event, None)
print('Response: %s\n' % json.dumps(response))
```

We open the sample GET file, parse the JSON into Dict, and then pass it as an argument to lambda_query_dynamo.lambda_handler(). As we have not mocked DynamoDB, it will query the table specified in the table_name = 'user-visits-sam' Lambda function. It will then capture the output response, which could look like the following:

```
Response: {"statusCode": "200", "body": "[{\"EventCount\": 3, \"EventDay\":
20171001, \"EventId\": \"324\"}, {\"EventCount\": 5, \"EventDay\":
20171002, \"EventId\": \"324\"}, {\"EventCount\": 4, \"EventDay\":
20171010, \"EventId\": \"324\"}, {\"EventCount\": 20, \"EventDay\":
20171012, \"EventId\": \"324\"}, {\"EventCount\": 10, \"EventDay\":
20171013, \"EventId\": \"324\"}, {\"EventCount\": 6, \"EventDay\":
20171014, \"EventId\": \"324\"}, {\"EventCount\": 6, \"EventDay\":
20171016, \"EventId\": \"324\"}, {\"EventCount\": 2, \"EventDay\":
20171017, \"EventId\": \"324\"}]", "headers": {"Content-Type":
"application/json", "Access-Control-Allow-Origin": "*"}}
```

The body is the same as we saw in the browser in Chapter 3, *Deploying Your Serverless Stack*. As such, you can debug different integration scenarios directly with real data and build more complete test suites as you step though the Lambda code with real data.

Integration testing using real test data

Now that we understand the real test data, we will look at how we can test a deployed Lambda function. First, you will need to install and set up the AWS CLI and configure the AWS Credentials as shown at the end of Chapter 1, *Serverless Microservices Architectures and Patterns*:

```
$ sudo pip3 sudo install awscli
$ aws configure
```

We will redeploy the serverless microservice stack we deployed in Chapter 3, *Deploying Your Serverless Stack*, so that we can test it. Use the following commands:

```
$ cd ./serverless-microservice-data-api/bash/apigateway-lambda-dynamodb
$ ./build-package-deploy-lambda-dynamo-data-api.sh
```

This will rebuild the Lambda ZIP package as the code if there have been any changes. Then it will package and deploy the code and SAM configuration. Finally, it will create the API Gateway, Lambda function, and DynamoDB table.

For testing, we will be using the AWS CLI, which can invoke all of the AWS-managed services. Here we are interested in the <code>lambda</code> (https://docs.aws. amazon.com/cli/latest/reference/lambda/index.html) and <code>apigateway</code> (https://docs.aws.amazon.com/cli/latest/reference/ apigateway/index.html) services.

Testing that a Lambda has been deployed correctly

To test a deployed Lambda, you can run the following command:

```
$ aws lambda invoke --invocation-type Event \
  --function-name lambda-dynamo-data-api-sam  --region eu-west-1 \
  --payload file://../../sample_data/request-api-gateway-get-valid.json \
outputfile.tmp
```

To automate it, we can put the following code into a shell script, serverless-microservice-data-api/bash/apigateway-lambda-dynamodb/invoke-lambda.sh:

```
#!/bin/sh
. ./common-variables.sh
rm outputfile.tmp
status_code=$(aws lambda invoke --invocation-type RequestResponse \
    --function-name ${template}-sam --region ${region} \
    --payload file://../../sample_data/request-api-gateway-get-valid.json \
    outputfile.tmp --profile ${profile})
echo "$status_code"
if echo "$status_code" | grep -q "200";
then
    cat outputfile.tmp
    if grep -q error outputfile.tmp;
    then
        echo "\nerror in response"
        exit 1
    else
        echo "\npass"
        exit 0
    fi
else
    echo "\nerror status not 200"
    exit 1
fi
```

We invoke the Lambda, but also check the response given in the `outputfile.tmp` file using the `grep` command. We return an exit code of 1 if an error is detected, and 0 otherwise. This allows you to chain logic when involved by other tools or by CI/CD steps.

Testing that API Gateway has been deployed correctly

We also want to be able to test that the serverless microservice API is working correctly after it is deployed. I use a mix of Python and bash to make it easier.

A Python script called `serverless-microservice-data-api/bash/apigateway-lambda-dynamodb/get_apigateway_endpoint.py` is first used to query AWS API Gateway to get the full endpoint and return a code 0 if it succeeds:

```python
import argparse
import logging

import boto3
logging.getLogger('botocore').setLevel(logging.CRITICAL)

logger = logging.getLogger(__name__)
logging.basicConfig(format='%(asctime)s %(levelname)s %(name)-15s:
%(lineno)d %(message)s',
                    level=logging.INFO) logger.setLevel(logging.INFO)


def get_apigateway_names(endpoint_name):
    client = boto3.client(service_name='apigateway',
                          region_name='eu-west-1')
    apis = client.get_rest_apis()
    for api in apis['items']:
        if api['name'] == endpoint_name:
            api_id = api['id']
            region = 'eu-west-1'
            stage = 'Prod'
            resource = 'visits/324'
            #return F"https://{api_id}.execute-api.
             {region}.amazonaws.com/{stage}/{resource}"
            return "https://%s.execute-api.%s.amazonaws.com/%s/%s"
                % (api_id, region, stage, resource)
    return None


def main():
```

```
    endpoint_name = "lambda-dynamo-xray"

    parser = argparse.ArgumentParser()
    parser.add_argument("-e", "--endpointname", type=str,
        required=False, help="Path to the endpoint_name")
    args = parser.parse_args()

    if (args.endpointname is not None): endpoint_name =
        args.endpointname

    apigateway_endpoint = get_apigateway_names(endpoint_name)
    if apigateway_endpoint is not None:
        print(apigateway_endpoint)
        return 0
    else:
        return 1

if __name__ == '__main__':
    main()
```

Then we use a shell script to call the Python script. The Python script returns the API endpoint, which is used in the curl with the sample GET request. We then look to see whether we get a valid status code.

Here is the full script for `serverless-microservice-data-api/bash/apigateway-lambda-dynamodb/curl-api-gateway.sh`:

```
. ./common-variables.sh
endpoint="$(python get_apigateway_endpoint.py -e ${template})"
echo ${endpoint}
status_code=$(curl -i -H \"Accept: application/json\" -H \"Content-Type:
application/json\" -X GET ${endpoint})
echo "$status_code"
if echo "$status_code" | grep -q "HTTP/1.1 200 OK";
then
    echo "pass"
    exit 0
else
    exit 1
fi
```

Having these scripts set up in this way allows us to easily automate these integrations tests.

Functions as a Service (FaaS) is still a relatively new area. There are still many discussions on the types of integration tests that should be used. One view is that we should do the full suite of testing in a different AWS account, especially the ones that would write or update a data store, such as POST or PUT requests.

I've included `--profile` and `aws_account_id` if you want to do this. In addition, with API Gateway, you can use a wide range of test suites that already exist around the HTTP endpoints, but testing other AWS services integration with Lambdas, such as objects being created in S3 that trigger a Lambda, needs a little bit more work and thought. In my view, serverless integration tests are still less mature, but I have already shown how they can be achieved by invoking the Lambda function directly with AWS CLI and Lambda with a JSON event source payload or invoking the API Gateway endpoint directly with a `curl` command.

Next we will look at how the SAM CLI can also be used for local testing.

The AWS Serverless Application Model CLI

In this section, we will walk through different features of SAM Local with fully-working examples. For local testing, you can use Python and bash like I have shown or you can also use the SAM CLI (`https://github.com/awslabs/aws-sam-cli`), which at the time of writing is still in beta. It uses Docker and is based on open source `docker-lambda` (`https://github.com/lambci/docker-lambda`) Docker images. If you are using Windows 10 Home, I recommend you upgrade to Pro or Enterprise as it's harder to get Docker working on the Home edition. There are also some hardware requirements, such as virtualization, to be aware of. We need to perform the following steps:

1. Install the AWS CLI (`https://docs.aws.amazon.com/cli/latest/userguide/installing.html`).
2. Install Docker CE (`https://docs.docker.com/install/`).
3. Install the AWS SAM CLI (`https://docs.aws.amazon.com/serverless-application-model/latest/developerguide/serverless-sam-cli-install.html`).
4. For Linux, you can run the following:

   ```
   $sudo pip install --user aws-sam-cli
   ```

5. For Windows, you can install AWS SAM CLI using an MSI.

6. Create a new SAM Python 3.6 project, `sam-app`, and `docker pull` the images (this should happen automatically but I needed to do `pull` to get it to work):

   ```
   $ sam init --runtime python3.6
   $ docker pull lambci/lambda-base:build
   $ docker pull lambci/lambda:python3.6
   ```

7. Invoke the following function:

```
$ cd sam-app
$ sam local invoke "HelloWorldFunction" -e event.json --region eu-
west-1
```

You will get the following:

```
Duration: 8 ms Billed Duration: 100 ms Memory Size: 128 MB Max
Memory Used: 19 MB
{"statusCode": 200, "body": "{\"message\": \"hello world\"}"}
```

This can be used to add automated testing.

8. Start the local Lambda endpoint:

```
$ sam local start-lambda --region eu-west-1
# in new shell window
$ aws lambda invoke --function-name "HelloWorldFunction" \
    --endpoint-url "http://127.0.0.1:3001" --no-verify-ssl out.txt
```

This starts a Docker container that emulates AWS Lambda with an HTTP server locally, which you can use to automate the testing of your Lambda functions from the AWS CLI or Boto3.

9. Start an API and test it using the following:

```
$ sam local start-api --region eu-west-1
# in new shell window
$ curl -i -H \"Accept: application/json\" -H \"Content-Type:
application/json\" -X GET http://127.0.0.1:3000/hello
```

This starts a Docker container with an HTTP server locally, which you can use to automate the testing of the API you can use with `curl`, Postman, or your web browser.

10. One way to generate sample events is to print out the event from a Lambda, and copy it from CloudWatch logs (my preference). Another way is to use `sam local`, which can generate some examples events. For example, you can run the following:

```
$ sam local generate-event apigateway aws-proxy
```

Personally, I haven't used the SAM CLI extensively as it is very new, needs Docker installed, and is still in beta. But it does look promising, and as another tool to test your serverless stack, it's useful that it can simulate a Lambda in Docker container that exposes an endpoint, and I expect more features to be added in the future.

Perhaps less usefully, it also wraps some of the existing command's serverless package and deployment commands as an alias for the CloudFormation ones. I think this is done to keep them all in one place.

11. Here is an example of the SAM CLI `package` and `deploy` commands:

```
$ sam package \
    --template-file $template.yaml \
    --output-template-file ../../package/$template-output.yaml \
    --s3-bucket
$bucket $ sam deploy \
    --template-file ../../package/$template-output.yaml \
    --stack-name $template \
    --capabilities CAPABILITY_IAM
```

CloudFormation with SAM to `package` and `deploy` commands:

```
$ aws cloudformation package \
    --template-file $template.yaml \
    --output-template-file ../../package/$template-output.yaml \
    --s3-bucket $bucket \
    --s3-prefix $prefix
$ aws cloudformation deploy \
    --template-file ../../package/$template-output.yaml \
    --stack-name $template \
    --capabilities CAPABILITY_IAM
```

Loading and end-to-end testing at scale

Next, we are going to take a look at Locust, which is a Python tool for performance and load testing. Then we are going to talk about strategies to reduce the API's latency and improve the response time of the API, and using Locust will show us the performance improvements.

Load testing your serverless microservice

First, you need to have a serverless microservice stack running with `./build-package-deploy-lambda-dynamo-data-api.sh`, and have loaded data into the DynamoDB table using the `python3 dynamo_insert_items_from_file.py` Python script.

Then install Locust, if it hasn't already been installed with the other packages in `requirements.txt`:

```
$ sudo pip3 install locustio
```

Locust (`https://docs.locust.io`) is an easy-to-use load-testing tool with a web metrics and monitoring interface. It allows you to define user behavior using Python code and can be used to simulate millions of users over multiple machines.

To use Locust, you first need to create a Locust Python file where you define the Locust tasks. The `HttpLocust` class adds a client attribute that is used to make the HTTP request. A `TaskSet` class defines a set of tasks that a Locust user will execute. The `@task` decorator declares the tasks for `TaskSet`:

```python
import random
from locust import HttpLocust, TaskSet, task

paths = ["/Prod/visits/324?startDate=20171014",
         "/Prod/visits/324",
         "/Prod/visits/320"]


class SimpleLocustTest(TaskSet):

    @task
    def get_something(self):
        index = random.randint(0, len(paths) - 1)
        self.client.get(paths[index])


class LocustTests(HttpLocust):
    task_set = SimpleLocustTest
```

To test the GET method with different resources and parameters, we are selecting three different paths randomly from a paths list, where one of the IDs does not exist in DynamoDB. The main idea is that we could easily scale this out to simulate millions of different queries if we had loaded their corresponding rows from a file into DynamoDB. Locust supports much more complex behaviors, including processing responses, simulating user logins, sequencing, and event hooks, but this script is a good start.

To run Locust, we need to get the API Gateway ID, which looks like `abcdefgh12`, to create the full hostname used for load testing. Here, I wrote a Python script called `serverless-microservice-data-api/bash/apigateway-lambda-dynamodbget_apigateway_id.py` that can do so based on the API name:

```python
import argparse
import logging

import boto3
logging.getLogger('botocore').setLevel(logging.CRITICAL)

logger = logging.getLogger(__name__)
logging.basicConfig(format='%(asctime)s %(levelname)s %(name)-15s:
%(lineno)d %(message)s',
                    level=logging.INFO)
logger.setLevel(logging.INFO)


def get_apigateway_id(endpoint_name):
    client = boto3.client(service_name='apigateway',
            region_name='eu-west-1')
    apis = client.get_rest_apis()
    for api in apis['items']:
        if api['name'] == endpoint_name:
            return api['id']
    return None

def main():
    endpoint_name = "lambda-dynamo-xray"

    parser = argparse.ArgumentParser()
    parser.add_argument("-e", "--endpointname", type=str,
                        required=False, help="Path to the endpoint_id")
    args = parser.parse_args()

    if (args.endpointname is not None): endpoint_name = args.endpointname


    apigateway_id = get_apigateway_id(endpoint_name)
    if apigateway_id is not None:
        print(apigateway_id)
        return 0
    else:
        return 1

if __name__ == '__main__':
    main()
```

Run the following commands to launch Locust:

```
$ . ./common-variables.sh
$ apiid="$(python3 get_apigateway_id.py -e ${template})"
$ locust -f ../../test/locust_test_api.py --
host=https://${apiid}.execute-api.${region}.amazonaws.com
```

Alternatively, I also have this `locust` run commands as a shell script you can run under the `test` folder `serverless-microservice-data-api/bash/apigateway-lambda-dynamodb/run_locus.sh`:

```
#!/bin/sh
. ./common-variables.sh
apiid="$(python3 get_apigateway_id.py -e ${template})"
locust -f ../../test/locust_test_api.py --
host=https://${apiid}.execute-api.${region}.amazonaws.com
```

You should now see Locust start in the Terminal and perform the following steps:

1. Navigate to `http://localhost:8089/` in your web browser to access the Locust web-monitoring and -testing interface.
2. In the **Start New Locust** swarm, enter the following:
   - 10 for **Number of users** to simulate
   - 5 for Hatch rate (users spawned/second)
3. Leave the tool running on the **Statistics** tab for a few minutes.

You will get something like this in the **Statistics** tab:

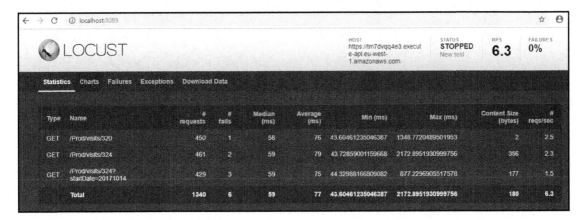

And on the **Charts** tab, you should get something similar to the following:

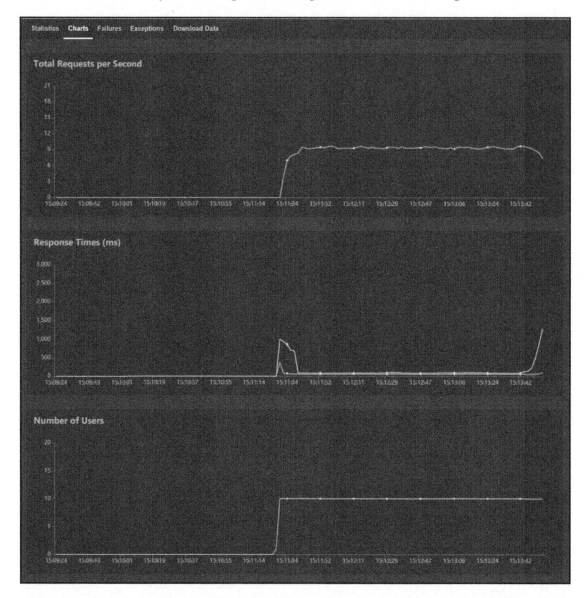

In the **Response Times (ms)** chart, the orange line represents the 95th percentile, and green is for the median response times.

Here are some observations about the preceding charts:

- The maximum request time is 2,172 milliseconds or about 2.1 seconds, which is really slow—this is linked to what is known as a cold start, which is the slower way to first launch a Lambda.
- The number of fails also goes up after about a minute—this is because DynamoDB permits some burst reads before it starts to throttle the read requests. If you log onto the AWS Management Console and look at the DynamoDB table metrics, you will see that this is happening:

Strategies to reduce the API's latency

There are many strategies to reduce latency. We will look at two, both of which are set in the SAM template:

- **Increasing the Lambda RAM size**: Currently, it is set to the minimum of 128 MB
- **Increasing the DynamoDB Read Capacity**: Currently, it is set to the smallest value of 1 unit

What I really like about DynamoDB is that you can change the capacity per table, and change the write capacity independently of the read capacity. This is very interesting and cost-effective for me in read-heavy use cases, where I can set the read capacity higher than the write capacity. There are even options to have the table autoscale based on the read/write utilization, or have the capacity purely based **on demand**, where you pay per read/write request.

We will start by increasing the DynamoDB read capacity for the table from 1 to 500 read units (keep the write capacity at 1 units). The cost was $0.66/month, but it will now increase to $55.24/month.

Edit the `lambda-dynamo-data-api.yaml` SAM YAML template file and increase `ReadCapacityUnits` from 1 to 500:

```
AWSTemplateFormatVersion: '2010-09-09'
Transform: 'AWS::Serverless-2016-10-31'
Description: >-
  This Lambda is invoked by API Gateway and queries DynamoDB.
Parameters:
    AccountId:
        Type: String

Resources:
  lambdadynamodataapi:
    Type: AWS::Serverless::Function
    Properties:
      Handler: lambda_return_dynamo_records.lambda_handler
      Runtime: python3.6
      CodeUri: ../../package/lambda-dynamo-data-api.zip
      FunctionName: lambda-dynamo-data-api-sam
      Description: >-
        This Lambda is invoked by API Gateway and queries DynamoDB.
      MemorySize: 128
      Timeout: 3
      Role: !Sub 'arn:aws:iam::${AccountId}:role/
```

```
                    lambda-dynamo-data-api'
        Environment:
          Variables:
            environment: dev
        Events:
          CatchAll:
            Type: Api
            Properties:
              Path: /visits/{resourceId}
              Method: GET
  DynamoDBTable:
    Type: AWS::DynamoDB::Table
    Properties:
      TableName: user-visits-sam
      SSESpecification:
        SSEEnabled: True
      AttributeDefinitions:
        - AttributeName: EventId
          AttributeType: S
        - AttributeName: EventDay
          AttributeType: N
      KeySchema:
        - AttributeName: EventId
          KeyType: HASH
        - AttributeName: EventDay
          KeyType: RANGE
      ProvisionedThroughput:
        ReadCapacityUnits: 500
        WriteCapacityUnits: 1
```

Run `./build-package-deploy-lambda-dynamo-data-api.sh` to deploy the serverless stack with the DynamoDB table changes.

Now run Locust again with 10 users with a hatch rate of 5:

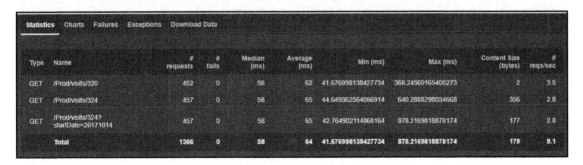

Type	Name	# requests	# fails	Median (ms)	Average (ms)	Min (ms)	Max (ms)	Content Size (bytes)	# reqs/sec
GET	/Prod/visits/320	452	0	58	62	41.676998138427734	368.24560165405273	2	3.5
GET	/Prod/visits/324	457	0	58	65	44.649362564086914	640.2888298034668	356	2.8
GET	/Prod/visits/324? startDate=20171014	457	0	58	65	42.764902114868164	878.2169818878174	177	2.8
	Total	1366	0	58	64	41.676998138427734	878.2169818878174	179	9.1

And on the **Charts** tab, you should get something similar to the following:

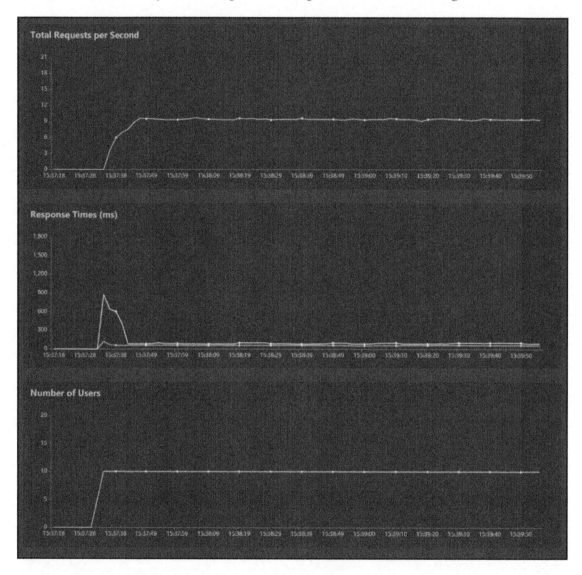

Here are some observations about the preceding charts:

- There are zero faults
- The average response time is 64 milliseconds, which is excellent

We get these results because the DynamoDB table read capacity was increased, that is, the requests are no longer getting throttled.

Now increase the RAM available in the Lambda function:

1. Modify the `lambda-dynamo-data-api.yaml` SAM YAML file by changing **MemorySize: 128** to **MemorySize: 1536.**
2. Run `./build-package-deploy-lambda-dynamo-data-api.sh` to deploy the serverless stack with the Lambda RAM changes.

Here are some observations that we've made the preceding changes:

- There are zero faults
- The average response time is 60 milliseconds, which is slightly better, especially considering this is a round trip of API Gateway to Lambda to DynamoDB and back

With 100 users with a hatch rate of 10, we get the following results:

- There are zero faults
- The average response time is 66 milliseconds, with a max of 1,057 milliseconds at the start of the load test

With 250 users with a hatch rate of 50, we get the following results:

- There are zero faults
- The average response time is 81 ms, with a max of 1,153 milliseconds at the start of the load test

You can test with a higher number of concurrent users as well, such as 1,000, and it will still work without faults even if the response time will be much higher to due to other bottlenecks. If you want to scale out further, I recommend you consider a different architecture. It is still impressive to have a serverless microservice scale up so easily, with just a couple of parameter changes in a config file!

This gives you a very good idea of how you can reduce the latency of your API.

Cleaning up

Run the following shell script with `./delete-stack.sh` to delete the serverless stack:

```bash
#!/usr/bin/env bash
. ./common-variables.sh
aws cloudformation delete-stack --stack-name $template --region $region --profile $profile
```

Summary

In this chapter, we explored many types of testing, including unit tests with mocks, integration testing with Lambda and API Gateway, debugging a Lambda locally, making a local endpoint available, and load testing. This is something that we will build upon in the rest of this book.

In the next chapter, we will look at serverless, distributed data management patterns and architectures that you can apply within your organization.

5
Securing Your Microservice

In this chapter, we will get a brief overview of security in AWS to ensure that your serverless microservices are secure. Before we create our first microservice, we first need to understand the AWS security models. We're going to discuss different terms that are important and the overall AWS security model. We are then going to talk about IAM, which is used to access any of the AWS resources. Finally, we'll look at securing your serverless microservice.

We will cover the following topics in this chapter:

- Overview of the security in AWS
- Overview of AWS Identity and Access Management (IAM)
- Securing your serverless microservice

Overview of the security in AWS

In this section, we are going to provide an overview of the security in AWS. We are going to take a look at why security is important, provide some examples of security, discuss the types of security terms that are important, and talk about the AWS shared-responsibility model.

Why is security important?

The following points discuss the importance of security:

- **Compliance with the law and standards**: For example, the European Union **General Data Protection Regulation (GDPR)** and **Health Insurance Portability and Accountability Act (HIPAA)** adopted by the US are responsible for regulating the laws for data protection and privacy for all individuals.

- **Data integrity**: Systems that aren't secure could have their data stripped or tampered with, which means that you can no longer trust the data.
- **Personally-identifiable information**: Privacy is a major concern these days. You should protect your user data and assets as a matter of course.
- **Data availability**: If an auditor, for example, asks you for specific data, you need to be able to retrieve that data. If a natural disaster happens near your data center, that data needs to be available and secure.

Let's have a look at the following list:

Insecure Communication	Social engineering	Litigation Costs
Misconfiguration	Viruses	Data Loss
Unpatched Vulnerabilities	Hacking	Reputation Costs
Insecure Disposal	Malware	Data Breaches
Missing Updates	Hacktivist	Financial Costs
Insecure Storage	Ransomware	Business Disruption
Leaked Keys	Rootkit	Report Incidents
Weak Physical Security	Phishing	Ransom Payments
	Cracking	
	Distributed Denial of Service Attack (DDoS)	

On the left, we have various systems that are configured incorrectly, missing updates, or have unsecured communication means. This could actually lead to the middle section, where there are problems such as the systems will get hacked or there'll be a ransomware demand or there could be an infiltration into your systems. A distributed denial service attack could be made, for example, which will bring down your e-commerce website so it will no longer be available to customers.

On the right, you can see some of the impacts. There could be litigation costs, data loss or data leaks, a financial cost to your organization, as well as reputational costs.

Types of security terms in AWS

A lot of the security in AWS is actually configuration and just having the correct architecture in place. So, it's important to understand some of these security terms:

- **Security and transit**: Think of this as HTTPS SSL. If you think about a web browser, you would have the padlock in your browser to show that the communication is secure, for example, when you're accessing any of your online banking systems.
- **Security at rest**: This is the data that's encrypted in the database or filesystem. Only a user with the key has access to the data.
- **Authentication**: This refers to the process to confirm whether a user or system is who they are meant to be.
- **Authorization**: Once you're authenticated, the system checks for correct authorization. This is to check that the permissions and access controls are in place for you to access the specific AWS resources.

Overview of AWS Identity and Access Management (IAM)

In this section, we are going to briefly discuss AWS IAM, specifically for serverless computing. IAM is a central location where you can manage users and security credentials—such as password, access keys, and permission policies—that control access to the AWS services and resources. We are going to talk about the most relevant IAM resources: policies, roles, groups, and users.

IAM policies are JSON documents that define the affected action's resources and conditions. Here is an example of a JSON document that will grant read access to DynamoDB tables, called `Books` only if the request originates from a specific IP range:

```
{
    "Version": "2012-10-17",
    "Statement": {
        "Effect": "Allow",
        "Action": [  "dynamodb:GetItem",
                    "dynamodb:Scan",
                    "dynamodb:Query"],
        "Resource": "arn: aws: dynamodb: eu-west-1: 123456789012: table/Books",
        "Condition": {
            "IpAddress": {
                "aws: SourceIp": "10.70.112.23/16"
            }
        }
    }
}
```

There is also a visual editor that allows you to create these or you can do so manually by editing the JSON document itself.

IAM users

An IAM user is a person or service used to interact with AWS. They access the Management Console via a password or multi-factor authentication (for the new user), or they may have an access key for programmatic access using the command-line interface or the SDKs. As shown in the following diagram, you can attach a policy to a user to grant them access to a resource to read DynamoDB within a specific IP range:

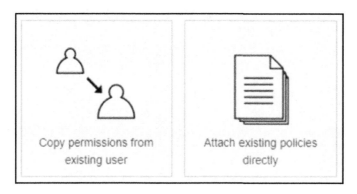

IAM groups

IAM groups are used to better mimic the security terms in your organization groups. You could think of them as Active Directory groups. For example, in your organization, you would have administrators, developers, and testers.

To create a group, you can use the AWS Management Console, SDK, or CLI under the IAM add group and then attach a policy. Once you have created a group, you can attach it to a user or you can create a new one.

IAM roles

IAM roles are similar to users, in that they can have a policy attached to them, but they can be attached by anyone who needs access in a trusted entity. In that way, you can delegate access to users, applications, or services without having to give them a new AWS key, as they could use the temporary security tokens through this trusted entity. For example, you could grant a third-party read access to an S3 bucket and nothing else within your AWS environment without actually having to share any keys and purely using the roles:

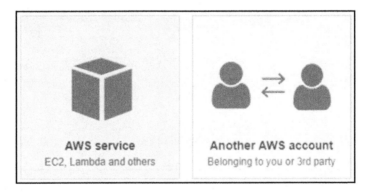

Securing your serverless microservice

In this section, we are going to talk about the security required to build your first microservice. Specifically, we are going to look at the security around Lambda functions, API Gateway, and DynamoDB, and then we are going to discuss the ways you can use monitoring and alerting upon detecting suspicious events.

Lambda security

In lambda security, there are two types of IAM roles:

- **Invoking the lambda**: This means having the permissions to actually invoke and run a lambda function. For example, this could be from an API Gateway or another service.
- **Granting lambda function read and write access to specific AWS resources**: For example, you would allow a Lambda function to read from a DynamoDB table.

In addition, the **Key Management Service (KMS)**, which is an AWS-managed service for keys, allows you to perform encryption and decryption on data at rest such as in a database or a NoSQL data store, such as DynamoDB. Amazon Virtual Private Cloud is another option where Lambda runs, by default, within a secure VPC. However, you may want to run it inside your own private AWS VPC if there are resources you need to access, such as elastic clash clusters or RDS, that are within this private VPC. Here is a work flow representation of using AWS Lambda using AWS KMS and AWS VPC:

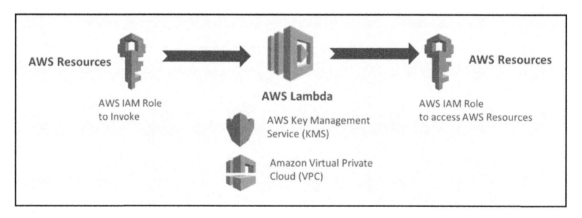

For API Gateway security, there are three ways you can control who can call your API method. This is known as request authorization, shown in the following diagram:

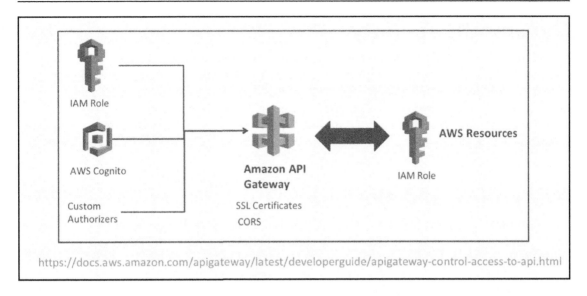

https://docs.aws.amazon.com/apigateway/latest/developerguide/apigateway-control-access-to-api.html

Here are the different ways to control who can call your API:

- **IAM roles and policies**: This provides access to the API Gateway. API Gateway will use these roles and policies to verify the caller's signature upon request.
- **Amazon Cognito user pools**: This controls who can access the API. In this case, the user will have to sign in to access the API.
- **An API Gateway custom authorizer**: This is a request, such as a bearer token or lambda function, that deals with validation and checks whether the client is authorized to call the API.

If you get requests from a domain other than your API's own domain, you must enable cross-origin resource sharing. In addition, API Gateway supports SSL certificates and certificate authorities. API Gateway may need authorization via an IAM role to call or invoke specific resources within AWS, such as with Kinesis streams or to invoke a Lambda function.

DynamoDB security

You can perform authentication using the IAM user or you can use a specific IAM role. Once they're authenticated, the authorization is controlled and the IAM policy is assigned to that specific user or role. What I recommend is that, when creating these policies for DynamoDB, you lock them down as much as possible, which means avoiding read and write access to all of the tables and DynamoDB. It's better to use a specific name for specific tables.

Monitoring and alerting

It's important to monitor systems for any suspicious activity and to detect any performance issues. API Gateway, DynamoDB, and Lambda functions all support CloudTrail, CloudWatch, and X-Ray for monitoring and alerting. They are discussed as follows:

- CloudTrail allows you to monitor all APIs and access to resources by any user or system.
- CloudWatch allows you to collect and track metrics and monitor log files, set specific alarms, and automatically react to changes in your AWS resources.
- X-Ray is a new service that traces requests and can generate service Maps.

The combination of these free systems gives you very good insight, out of the box, into your serverless system.

Summary

After reading this chapter, you should have a much deeper understanding of security in AWS and why it's important for your organization. After all, no one wants to be the person responsible for a data breach. We discussed IAM and you now know that policies are the key documents that ensure restricted access to AWS resources. We also looked at some of the security concepts that secure your serverless microservices; specifically, we learned about lambda, API Gateway, and DynamoDB.

Other Books You May Enjoy

If you enjoyed this book, you may be interested in these other books by Packt:

Cloud Native Architectures
Tom Laszewski, Kamal Arora, Et al

ISBN: 978-1-78728-054-0

- Learn the difference between cloud native and traditional architecture
- Explore the aspects of migration, when and why to use it
- Identify the elements to consider when selecting a technology for your architecture
- Automate security controls and configuration management
- Use infrastructure as code and CICD pipelines to run environments in a sustainable manner
- Understand the management and monitoring capabilities for AWS cloud native application architectures

Serverless Programming Cookbook
Heartin Kanikathottu

ISBN: 978-1-78862-379-7

- Serverless computing in AWS and explore services with other clouds
- Develop full-stack apps with API Gateway, Cognito, Lambda and DynamoDB
- Web hosting with S3, CloudFront, Route 53 and AWS Certificate Manager
- SQS and SNS for effective communication between microservices
- Monitoring and troubleshooting with CloudWatch logs and metrics
- Explore Kinesis Streams, Amazon ML models and Alexa Skills Kit

Leave a review - let other readers know what you think

Please share your thoughts on this book with others by leaving a review on the site that you bought it from. If you purchased the book from Amazon, please leave us an honest review on this book's Amazon page. This is vital so that other potential readers can see and use your unbiased opinion to make purchasing decisions, we can understand what our customers think about our products, and our authors can see your feedback on the title that they have worked with Packt to create. It will only take a few minutes of your time, but is valuable to other potential customers, our authors, and Packt. Thank you!

Index

Printed in Great Britain
by Amazon

44906990R00097